Power Electronics and Power Systems

Series Editor

Joe H. Chow
Alex M. Stankovic
David Hill

More information about this series at http://www.springer.com/series/6403

Elias Kyriakides • Siddharth Suryanarayanan •
Vijay Vittal
Editors

Electric Power Engineering Research and Education

A festschrift for Gerald T. Heydt

Springer

Editors
Elias Kyriakides
KIOS Research Center for
Intelligent Systems
and Networks
and Department of Electrical
and Computer Engineering
University of Cyprus
Nicosia, Cyprus

Siddharth Suryanarayanan
Department of Electrical
and Computer Engineering
Colorado State University
Fort Collins, CO, USA

Vijay Vittal
Electrical, Computer and Energy Engineering
Arizona State University
Tempe, AZ, USA

ISSN 2196-3185 ISSN 2196-3193 (electronic)
Power Electronics and Power Systems
ISBN 978-3-319-35307-4 ISBN 978-3-319-17190-6 (eBook)
DOI 10.1007/978-3-319-17190-6

Springer Cham Heidelberg New York Dordrecht London

Softcover reprint of the hardcover 1st edition 2015

Printed on acid-free paper

Springer International Publishing AG Switzerland is part of Springer Science+Business Media (www.springer.com)

Foreword

This Festschrift documents a portion of the contributions of Gerald Thomas Heydt to the electrical engineering profession and, specifically, the field of electric power engineering. These contributions are as unique as the man himself, who is known for professionalism, innovation, responsiveness, and so many more attributes that continue to emerge as time goes on. On this occasion of Jerry's 70th birth year, the authors of this Festschrift have provided clear highlights of his work in harmonics and power quality, advanced control of energy and power systems, new T&D technologies, and most importantly his education of students. His pioneering work on stochastic methods is surfacing now in the context of uncertainty due to renewable resources, hybrid vehicles, and demand participation.

This strong collection of ideas and accomplishments lacks depth in one significant topic—service. Jerry Heydt performs professional service with great vigor and dedication. He knows that paper reviews, proposal reviews, committee participation, letter writing, and leadership are critical to the growth and prosperity of a profession. In addition, he has made it clear to the young people in our profession that service is a key factor in all of the other aspects of our work including education and research. His service to his students, fellow faculty, and our profession is nothing short of spectacular in every dimension.

Peter W. Sauer
W. W. Grainger Chair Professor in Electrical Engineering
University of Illinois at Urbana-Champaign
Champaign, IL, USA

Preface

It is a great honor and pleasure to present this Festschrift honoring Regents' Professor Gerald Thomas Heydt on the occasion of his 70th birthday. Jerry Heydt is a colleague, teacher, mentor, and above all a friend. He has made pioneering contributions in the area of electric power systems. His main technical achievements and accolades span three key areas: electric power quality, electric power transmission, and electric distribution networks. Over the last four decades, he has made lasting contributions in power engineering research and education that have profoundly impacted the field.

Gerald Thomas Heydt was born in 1943 in New York City. He graduated from the Bronx High School of Science in 1960. Prof. Heydt obtained his Bachelor of Engineering in Electrical Engineering from the Cooper Union in New York—a highly selective institution of higher education that offered full tuition scholarship to its students—in 1965. He then earned his Master of Science in Electrical Engineering and the Doctor of Philosophy from Purdue University in West Lafayette, Indiana, in 1967 and 1970, respectively. After his Ph.D, he joined the faculty of Purdue University, where he became Professor in 1980. During this period, he also worked briefly for the Commonwealth Edison in Chicago, Illinois. In 1990, he served as the Program Manager of the Power System program at the National Science Foundation in Washington DC. In 1994, he moved to Arizona State University (ASU) in Tempe, Arizona, where he also became the Site Director of the NSF Center for the Power Systems Engineering Research Center (PSERC). In 2002, Prof. Heydt was named Regents' Professor at ASU, the highest professorial rank in the University, recognizing both his technical and educational contributions to the University and the society. In 2009, he also became the Site Director of a National Science Foundation engineering research center called the Future Renewable Electric Energy Distribution and Management Center.

Professor Heydt's pioneering work in the area of power quality revolutionized the thinking of the power industry and led utilities and organizations to focus on the quality of the electric power supplied to the customer and to seek ways to alleviate the numerous problems in this area. His book on power quality was the only resource on this topic for a long time and is now the most cited resource in power quality. Today, power quality is a "hot" topic and a tremendous research effort around the globe is directed in this area. Gerald Heydt is credited with pioneering the initial research efforts in power quality research.

Numerous organizations have recognized Professor Heydt for his academic and research work. In 1997, Heydt was elected to the US National Academy of Engineering (NAE) for "contributions to the technology of electric power quality." Election to NAE is considered the highest distinction conferred to an engineer. In 1991, he was elected a Fellow of the Institute of Electrical and Electronics Engineers (IEEE) "For leadership in electric power engineering education and research on harmonic signals in electric power systems."

Professor Heydt is a passionate educator. He played a key role in resurrecting the activities of the IEEE Power and Energy Society (PES) Power Engineering Education Committee (now renamed the Power and Energy Education Committee). He is also an outstanding mentor and advisor. His teaching philosophy and contributions to power engineering education led the IEEE Power and Energy Society to recognize him with the IEEE PES Outstanding Power Engineering Educator award in 1995. Ethics and honesty in research and in his life are some of his distinguishing characteristics. These values are instilled in his students, colleagues, and collaborators.

This book is the result of a very successful event organized on Sunday, October 13, 2013 at Arizona State University: the Gerald T. Heydt Festschrift Symposium. The symposium included presentations from leading researchers whom Professor Heydt has mentored and collaborated with over the past four decades in power quality and electrical power systems. During this symposium, organized in conjunction with his 70th birthday, Professor Heydt was honored for his four decades of industry-changing innovation, scholarship, mentoring, and teaching in electric power engineering.

The book aims at presenting key advances in the three research areas where Prof. Heydt has had outstanding contributions. Some of the chapters describe work that Prof. Heydt was directly involved with, while other chapters are motivated by his research. Peter W. Sauer, the W. W. Grainger Professor of Electrical Engineering at the University of Illinois at Urbana-Champaign and fellow member of the NAE, introduces the book in the foreword. Prof. Sauer is a longtime collaborator and one of the first Ph.D. graduates of Professor Heydt.

The book comprises nine chapters. Chapters 1 and 2 focus on Power Quality. Chapter 1 introduces power system harmonics and discusses the injection of harmonic currents from nonlinear loads into distribution systems. Methods for harmonic analysis and algorithms for harmonic power flow study are also discussed. Chapter 2 describes a classification tool, based on a meta-heuristic algorithm, that is used for the classification of power quality disturbances.

Chapters 3–6 describe work in Transmission Engineering. Chapter 3 provides the experiences from five different projects that were collaboratively completed by Salt River Project and the team of Prof. Heydt at ASU. The projects investigated the application of synchrophasor technology in several aspects of wide area monitoring of power systems and in the parameter identification of synchronous machines. Chapter 4 introduces the reliability and availability analysis of wind and solar generation. Chapter 5 demonstrates the use of the geographical information system coordinates to aid in the calculation of unscheduled flows in wide-area power grids. The impact of wind variability on unscheduled flows is also investigated. Chapter 6 investigates the transmission expansion planning in modern power systems. The process for this complex decision-making process is outlined and mathematical models are described.

Chapter 7 provides an overview of the progress of automation in distribution systems and discusses several contemporary issues that are relevant for the modern distribution systems. This chapter also presents a detailed discussion of distribution automation functions, together with a cost-to-benefit analysis.

Chapters 8 and 9 focus on research, education, service, and workforce development in power engineering. Chapter 8 gives a personal account of the impact of Prof. Heydt to research, education, and service, as well as of his impact to the lives and careers of his colleagues. Chapter 9 provides the results of a regional labor market and workforce study that concentrated on electric power employers. The study aimed at looking into issues such as the aging utility workforce, retirements, and challenging population trends and made relevant conclusions and recommendations for the labor and skill gaps in the electric power industry.

It was our honor and privilege to organize the Gerald T. Heydt Festschrift Symposium in 2013. We hope the readers of this festschrift will find the chapters useful for their research and education.

Sincerely,
Elias Kyriakides, Siddharth Suryanarayanan, and Vijay Vittal

Memories from the Symposium

Prof. Heydt with attendees of the symposium

The night before the symposium at Frank Kush Field for a Sun Devil's game

Prof. Heydt with some of the symposium attendees

Contents

Chapter 1
Power System Harmonics

Surya Santoso and Anamika Dubey

1.1 Introduction

Electric power quality defines and quantifies the characteristics and quality of electric service. The specific definition of power quality can vary among different utilities and customers, therefore, there is no universal agreement over the definition of power quality and its scope. In general, electric power quality is measured and quantified as a function of deviation in the rated magnitude or electrical frequency of the load voltage or supply current waveforms. Therefore, an electric supply is said to be of poor power quality if the load voltage or supply current deviates from its rated magnitude and waveshape, or if the frequency composition of the sinusoidal voltage or current waveform changes. This broad definition also includes power system outages, which is generally a reliability concern in transmission and distribution operations. However, a momentary outage caused by the operation of overcurrent protection devices in clearing temporary faults is considered as a power quality issue.

Since a major part of power system engineering is concerned with the improvement of the quality of power supply, the term power quality can encompass topics ranging from traditional transmission and distribution engineering (e.g., substation design, grounding, etc.) to power system protection and power system planning. However, generally power quality phenomenon refers to the measurement, analysis, and improvement of voltage and current waveforms, so as to maintain the power supply (voltage and current) as a sinusoidal waveform at rated magnitude and frequency. This definition includes all momentary

S. Santoso (✉) • A. Dubey
Electrical and Computer Engineering, The University of Texas at Austin,
1616 Guadalupe, C0803, Austin, TX 78701, USA
e-mail: ssantoso@mail.utexas.edu

E. Kyriakides et al. (eds.), *Electric Power Engineering Research and Education*,
Power Electronics and Power Systems, DOI 10.1007/978-3-319-17190-6_1

phenomenon leading to problems in regulation and frequency of the power supply. For the scope of this chapter, we will restrict our discussion to the problems primarily relating to bus voltage and current waveshape and frequency characteristics. Several nonlinear loads such as rectifiers, adjustable-speed drives (ASD), and fluorescent lamps inject nonlinear current into the distribution circuit and may distort the sinusoidal load voltage and supply current waveforms. The distortions in the waveshape of the electric supply are measured in terms of harmonics. The harmonic components, which are integral multiples of the fundamental frequency, injected by nonlinear loads, modify the frequency characteristics of the power supply.

The objective of this chapter is to introduce the topic of power system harmonics and to discuss approaches used to analyze the distribution circuit in the presence of nonlinear loads. The chapter begins with an introduction on power system harmonics, harmonic sources, and their effects on the distribution system. Next, power system quantities are defined under non-sinusoidal operating conditions and commonly used power system indices used to measure harmonic distortions are defined. The characteristics of nonlinear loads injecting harmonic currents into the distribution system are discussed next. Because network components are generally modeled at the fundamental power frequency, their operational characteristics and circuit models are developed under non-sinusoidal conditions. Next, methods for harmonic analysis and algorithms for harmonic power flow study are discussed. Finally, harmonic filters used to mitigate harmonic concerns in the distribution grid are discussed. Note that the material for this chapter is primarily taken from the book Electric Power Quality written by Dr. Heydt [1]. He has done seminal work in the area of power quality, particularly in power system harmonics and harmonic power flow study [1–8]. His algorithms for harmonic power flow study were some of the first in this field. This chapter is a humble tribute to his valuable contribution in the area of power quality and power system harmonics.

1.2 Harmonics in Power Systems

This section introduces the problem of harmonics in power systems and their effects on distribution system power quality. Several power system quantities that measure and quantify power system harmonics are also discussed. Power system harmonics are typically introduced into the distribution system in the form of currents whose frequencies are the integral multiples of the fundamental power system frequency. These currents are produced by nonlinear loads, such as arc furnaces, rectifiers, fluorescent lamps, and electronic devices, which may distort the voltage and current waveforms. A high level of power system harmonics may lead to serious power quality problems.

In general, the power quality problems associated with harmonic distortions are caused by current distortions produced by nonlinear loads and thus originate at the customer load locations. The distorted current, which is also referred to as harmonic current, then interacts with the utility supply system impedance causing distortions in the load voltage and current, thus adversely affecting other users connected to the distribution system. Therefore, the nonlinear loads present in the distribution system result in a non-sinusoidal and periodic load current, which on interaction with the system impedance results in a non-sinusoidal periodic load voltage. Both non-sinusoidal voltage and current can be expressed as a weighted sum of sinusoids, whose frequencies are integral multiples of the fundamental frequency. This expression is called Fourier series expansion and the higher order frequency terms in the Fourier series are termed harmonics. Note that these harmonics are integral multiples of the fundamental frequency.

Let, $i(t)$, be the distorted current waveform of time period T. The current waveform is expressed as a Fourier series, given by (1.1).

$$i(t) = a_0 + \sum_{n=1}^{\infty} [a_n \cos(n\omega_1 t) + b_n \sin(n\omega_1 t)], \tag{1.1}$$

where $\omega_1 = \frac{2\pi}{T}$, and n is the harmonic order.

$$\begin{aligned} a_0 &= \frac{1}{T}\int_0^T i(t)dt \\ a_n &= \frac{2}{T}\int_0^T i(t)\cos(n\omega_1 t)dt \\ b_n &= \frac{2}{T}\int_0^T i(t)\sin(n\omega_1 t)dt. \end{aligned} \tag{1.2}$$

In general, even order harmonics do not exist in a three-phase power system. The even order harmonics only originate when a current waveform is asymmetrical along the time axis. Since loads in the power system mostly inject symmetrical current, except for few nonlinear single-phase loads, such as rectifiers and fluorescent lamps, even order harmonic components are significantly smaller than the odd order harmonic components. Therefore, for power system harmonic analysis, the distorted current waveform can be expressed as a weighted sum of only odd order harmonics (1.3).

$$i(t) = I_1 \sin(\omega_1 t) + \sum_{n=3,5,7,\ldots}^{\infty} I_n \sin(n\omega_1 t). \tag{1.3}$$

Next, we will discuss power system quantities used for harmonic analysis. The power system quantities, such as root-mean-square (rms) value, apparent power, and power factor, are originally defined at the fundamental frequency. In the presence of the nonlinear loads, with non-sinusoidal load voltages and currents, the power system quantities need to be redefined. In the following section, important power system quantities are redefined under non-sinusoidal conditions [9].

1.2.1 Root-Mean-Square

For a sinusoidal current waveform, $i(t) = I_1 \sin(\omega_1 t + \theta_1)$, the rms value is given in (1.4), where I_{rms} is the rms value of the current waveform.

$$I_{\text{rms}} = \frac{1}{\sqrt{2}} I_1. \tag{1.4}$$

Under non-sinusoidal conditions, the current waveform can be expressed using the Fourier series expansion, given by (1.5).

$$i(t) = I_1 \sin(\omega_1 t + \theta_1) + \sum_{n=3,5,7,\ldots}^{\infty} I_n \sin(n\omega_1 t + \theta_n). \tag{1.5}$$

The rms value of the current waveform in (1.5) is given by (1.6).

$$\begin{aligned} I_{\text{rms}} &= \sqrt{\frac{1}{2} \sum_{n=1,3,5,\ldots}^{\infty} I_n^2} \\ &= \sqrt{\left[\frac{I_1}{\sqrt{2}}\right]^2 + \left[\frac{I_3}{\sqrt{2}}\right]^2 + \left[\frac{I_5}{\sqrt{2}}\right]^2 + \cdots} \\ &= \sqrt{\frac{I_1^2}{2} + \frac{1}{2} \sum_{n=3,5,7,\ldots}^{\infty} I_n^2} \\ &= \sqrt{I_{\text{rms}_1}^2 + I_{\text{rms}_H}^2}. \end{aligned} \tag{1.6}$$

Similarly, rms voltage under non-sinusoidal conditions is given by (1.7).

$$\begin{aligned} V_{\text{rms}} &= \sqrt{\frac{V_1^2}{2} + \frac{1}{2} \sum_{n=3,5,7,\ldots}^{\infty} V_n^2} \\ &= \sqrt{V_{\text{rms}_1}^2 + V_{\text{rms}_H}^2} \end{aligned} \tag{1.7}$$

1.2.2 Power in Non-sinusoidal Conditions

Next, a brief review of the active and reactive power in the presence of harmonics is presented. Consider a voltage $v(t)$ and current $i(t)$ expressed in terms of their harmonic components:

$$\begin{aligned} v(t) &= V_1 \sin(\omega_1 t + \phi_1) + \sum_{n=3,5,7,\ldots}^{\infty} V_n \sin(n\omega_1 t + \phi_n) \\ i(t) &= I_1 \sin(\omega_1 t + \theta_1) + \sum_{n=3,5,7,\ldots}^{\infty} I_n \sin(n\omega_1 t + \theta_n). \end{aligned} \tag{1.8}$$

Apparent Power (S) is defined as

$$S = V_{\mathrm{rms}} I_{\mathrm{rms}}. \tag{1.9}$$

Using the expression for rms current and voltage as defined in (1.6) and (1.7), S is given by (1.10).

$$\begin{aligned} S^2 &= V_{\mathrm{rms}}^2 I_{\mathrm{rms}}^2 \\ &= \left[V_{\mathrm{rms}_1}^2 + V_{\mathrm{rms}_H}^2\right]\left[I_{\mathrm{rms}_1}^2 + I_{\mathrm{rms}_H}^2\right]. \end{aligned} \tag{1.10}$$

On expanding the expression for the apparent power (1.10), we obtain

$$\begin{aligned} S^2 &= V_{\mathrm{rms}}^2 I_{\mathrm{rms}}^2 \\ &= V_{\mathrm{rms}_1}^2 I_{\mathrm{rms}_1}^2 + V_{\mathrm{rms}_1}^2 I_{\mathrm{rms}_H}^2 + V_{\mathrm{rms}_H}^2 I_{\mathrm{rms}_1}^2 + V_{\mathrm{rms}_H}^2 I_{\mathrm{rms}_H}^2. \end{aligned} \tag{1.11}$$

Next, using the expression derived in (1.11), several power expressions are defined under non-sinusoidal operating condition.

Fundamental Apparent Power (S_1) is defined as

$$S_1 = V_{\mathrm{rms}_1} I_{\mathrm{rms}_1}. \tag{1.12}$$

And fundamental active power (P_1) and reactive power (Q_1) are given by

$$\begin{aligned} P_1 &= S_1 \cos(\phi_1 - \theta_1) \\ Q_1 &= S_1 \sin(\phi_1 - \theta_1). \end{aligned} \tag{1.13}$$

Non-fundamental Apparent Power (S_{N}) is observed in the system due to the interaction of current and voltage distortions. The expression of non-fundamental apparent power (S_{N}) is given as follows

$$\begin{aligned} S_N^2 &= V_{\text{rms}_1}^2 I_{\text{rms}_H}^2 + V_{\text{rms}_H}^2 I_{\text{rms}_1}^2 + V_{\text{rms}_H}^2 I_{\text{rms}_H}^2 \\ &= S_{CDP}^2 + S_{VDP}^2 + S_H^2, \end{aligned} \tag{1.14}$$

where

$$\begin{aligned} S_{CDP} &= \text{Current distortion power} \\ S_{VDP} &= \text{Voltage distortion power} \\ S_H &= \text{Harmonic apparent power.} \end{aligned} \tag{1.15}$$

Current Distortion Power (S_{CDP}) is observed due to the interaction of the harmonic component of current with the fundamental voltage component. As shown in (1.16), S_{CDP} is expressed as the product of rms value of harmonic current and fundamental voltage component.

$$\begin{aligned} S_{CDP} &= V_{\text{rms}_1} I_{\text{rms}_H} \\ &= V_{\text{rms}_1} \times \sqrt{\sum_{n=3,5,7,\ldots}^{\infty} I_{\text{rms}_n}^2}. \end{aligned} \tag{1.16}$$

Similarly, **Voltage Distortion Power** (S_{VDP}) is defined as the product of rms value of harmonic voltage and the rms value of fundamental current (1.17).

$$\begin{aligned} S_{VDP} &= V_{\text{rms}_H} I_{\text{rms}_1} \\ &= \left[\sqrt{\sum_{n=3,5,7,\ldots}^{\infty} V_{\text{rms}_n}^2} \right] \times I_{\text{rms}_1}. \end{aligned} \tag{1.17}$$

Harmonic Apparent Power (S_{H}) is defined as the product of harmonic components of non-sinusoidal voltage and current (1.18).

$$\begin{aligned} S_H &= V_{\text{rms}_H} I_{\text{rms}_H} \\ &= \left[\sqrt{\sum_{n=3,5,7,\ldots}^{\infty} V_{\text{rms}_n}^2} \right] \times \left[\sqrt{\sum_{n=3,5,7,\ldots}^{\infty} I_{\text{rms}_n}^2} \right]. \end{aligned} \tag{1.18}$$

Using the expression for S_{H}, total harmonic active power P_{H} and total harmonic nonactive power N_{H} are defined (1.19).

$$\begin{aligned} P_H &= \sum_{n=3,5,7,\ldots}^{\infty} P_n \\ &= \sum_{n=3,5,7,\ldots}^{\infty} V_{\text{rms}_n} I_{\text{rms}_n} \cos(\phi_n - \theta_n) \\ N_H &= \pm\sqrt{S_H^2 - P_H^2}. \end{aligned} \tag{1.19}$$

1.2.3 Power Factor

For the sinusoidal case, the power factor is given by

$$\mathrm{PF} = \frac{P_1}{S_1}, \tag{1.20}$$

where P_1 and S_1 are fundamental active and apparent power, respectively.

Under non-sinusoidal conditions, the power factor takes both fundamental and harmonic power into account. Therefore, the power factor with current and voltage harmonics in the system is defined as the ratio of the total active power at all power frequencies and the total apparent power (1.21).

$$\mathrm{PF} = \frac{P}{S} = \frac{P_1 + P_H}{S}, \tag{1.21}$$

where P_1 is the fundamental active power, P_H is the total harmonic active power, and S is the total apparent power.

1.2.4 Total Harmonic Distortion

Total harmonic distortion (THD) is one of the classical power quality indices used to measure distortions in current and voltage waveforms. For a periodic waveform of period $T = 2\pi/\omega$, the THD of the waveform is defined as the ratio of rms value of the harmonics components (V_{rms_H}) and the fundamental component (V_{rms_1}). The current and voltage THD are defined in (1.22).

$$\begin{aligned} \mathrm{THD}_V &= \frac{V_{\mathrm{rms}_H}}{V_{\mathrm{rms}_1}} = \frac{\sqrt{\sum_{n=2}^{\infty} V_{\mathrm{rms}_n}^2}}{V_{\mathrm{rms}_1}} \\ \mathrm{THD}_I &= \frac{I_{\mathrm{rms}_H}}{I_{\mathrm{rms}_1}} = \frac{\sqrt{\sum_{n=2}^{\infty} I_{\mathrm{rms}_n}^2}}{I_{\mathrm{rms}_1}} \end{aligned} \tag{1.22}$$

THD is generally expressed as a percentage. The THD of either the voltage or the current waveform can be calculated to analyze the harmonic distortion in the quantity under consideration. The THD can be readily calculated to quantify harmonic distortion, however, the information of the full frequency spectrum is lost.

1.2.5 Total Demand Distortion

The total demand distortion (TDD) index is a measure of current distortion. TDD is defined as the ratio of rms value of harmonic component of the current waveform and the maximum load current (I_L).

$$\mathrm{TDD} = \frac{I_{\mathrm{rms}_H}}{I_L} = \frac{\sqrt{\sum_{n=2}^{\infty} I_{\mathrm{rms}_n}^2}}{I_L}, \tag{1.23}$$

where the maximum load current is defined as

$$I_L = \frac{P_D}{\sqrt{3} \times \mathrm{PF} \times kV_{LL}}. \tag{1.24}$$

Here, P_{D} is the average peak load demand measured over a year, PF is the average billed power factor, and kV_{LL} is line-to-line voltage measured at the load.

1.2.6 Distortion Index

The distortion index (DIN) is a commonly used standard to quantify the voltage and current harmonic distortions, outside North America. It is defined as follows:

$$\mathrm{DIN} = \frac{V_{\mathrm{rms}_H}}{V_{\mathrm{rms}}} = \frac{\sqrt{\sum_{n=2}^{\infty} (V_{\mathrm{rms}_n})^2}}{\sqrt{\sum_{n=1}^{\infty} (V_{\mathrm{rms}_n})^2}}. \tag{1.25}$$

The relationship between DIN and THD is given as in (1.26).

$$\mathrm{DIN} = \frac{\mathrm{THD}_V}{\sqrt{1 + \mathrm{THD}_V^2}}. \tag{1.26}$$

1.3 Sources of Harmonics

The harmonic distortions occur at load buses connected to nonlinear loads. These buses are modeled as sources of harmonic signals, injecting harmonic currents into the distribution circuit. The nature of the harmonic signal depends upon the type of nonlinearity introduced by the load. Thus, a fluorescent lamp would not have the same voltage–current-frequency characteristics as a rectifier. This section presents

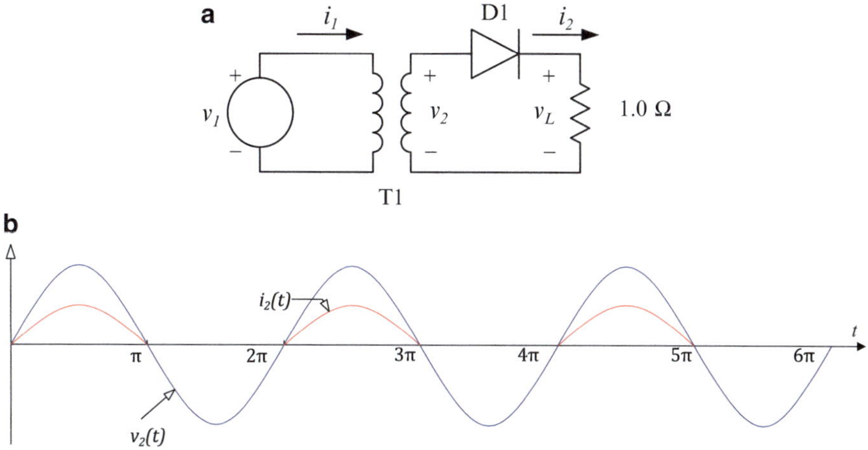

Fig. 1.1 (**a**) Simple ideal single-phase rectifier. (**b**) Corresponding voltage and current waveforms on the DC side

a review on the devices likely to induce harmonic distortions into the power system, and the characteristics of the harmonic currents produced by such devices.

1.3.1 Single- and Three-Phase AC/DC Power Converters

Power converters are devices capable of converting electrical energy from one frequency to another, typically from AC to DC or vice versa. Rectifiers and inverters convert AC to DC and DC to AC, respectively. Figure 1.1a shows a simple single-phase rectifier. Let transformer T1 and diode D1 be ideal elements. Diode D1 will conduct when

$$v_2(t) > 0. \tag{1.27}$$

The current and voltage waveforms $i_2(t)$ and $v_2(t)$ generated by the single-phase rectifier are shown in Fig. 1.1b. The Fourier series for the voltage and current waveforms produced by the single-phase rectifier shown in Fig. 1.1a are given by (1.28) and (1.29).

$$v_2(t) = \sin(t). \tag{1.28}$$

$$i_2(t) = \frac{1}{\pi} + \frac{1}{2}\sin(t) - \frac{2}{\pi}\sum_{n=1}^{\infty}\frac{1}{4n^2 - 1}\cos(2nt). \tag{1.29}$$

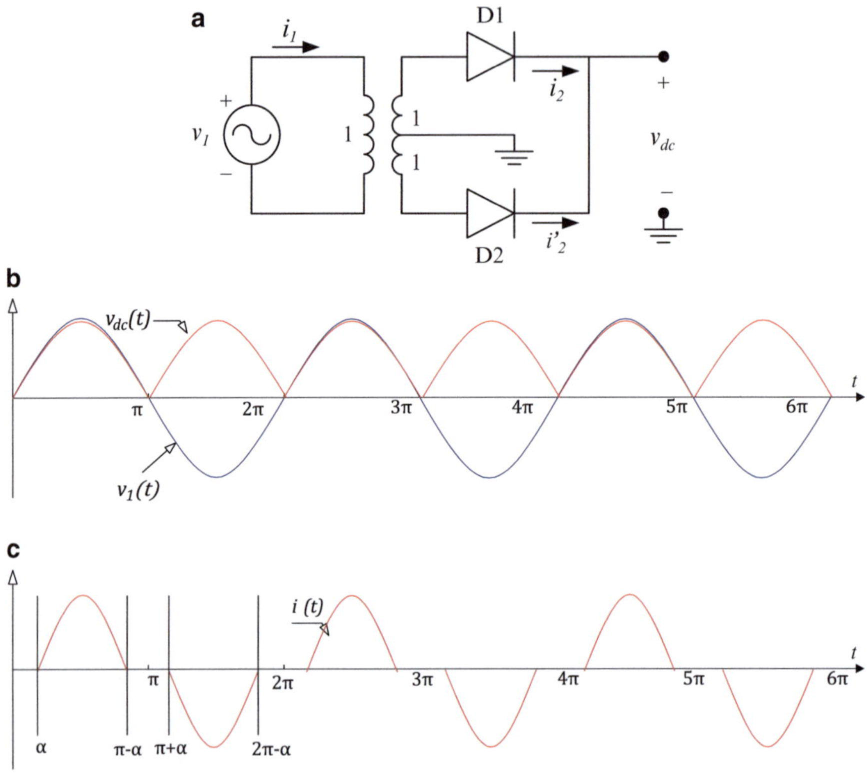

Fig. 1.2 (**a**) Half-wave single-phase rectifier. (**b**) DC circuit voltage for the half-wave single-phase rectifier. (**c**) Primary current for the half-wave rectifier

Next, Fig. 1.2a shows a more complex single-phase rectifier known as a half-wave rectifier. This circuit had been widely used in radio receivers and other communication equipment. This configuration is superior to the rectifier shown in Fig. 1.1a, as it does not produce any DC in the average magnetic flux produced by the secondary winding. Therefore, the transformer does not get biased by the DC value and the core saturation is avoided. Secondly, this configuration results in a smoother DC waveform as shown in Fig. 1.2b, thus facilitating easy filtering.

If the rectifier load is purely resistive, diode D1 will be on for $0 < t < \pi$, and diode D2 will be on for $\pi < t < 2\pi$. If no filtering is used, current i_1 will be sinusoidal. If filtering is used, the resulting current is shown in Fig. 1.2c. The value of α will depend upon the RC time constant of the filter. If C is too large, α will approach 2π; if C is too small, α will be closer to zero. Furthermore, current $i_1(t)$ can be expressed as the Fourier series given in (1.30).

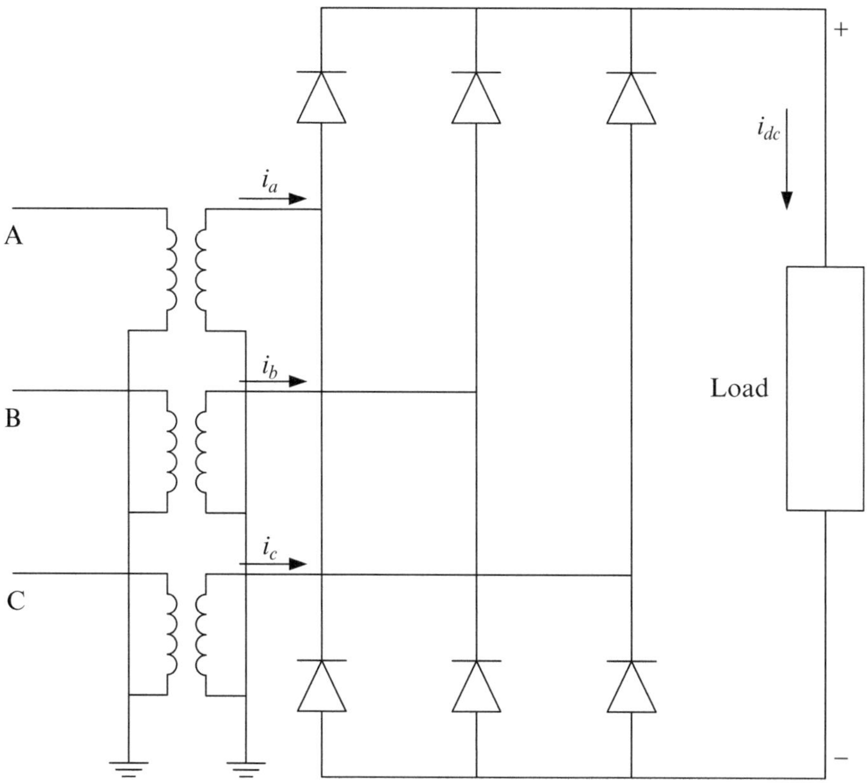

Fig. 1.3 Three-phase six-pulse power converter

$$i_1(t) = \frac{1}{\pi}[\pi - 2\alpha - \sin 2\alpha] \sin t - \frac{1}{6\pi}[2 \sin 2\alpha + \sin 4\alpha] \sin 3t + \cdots . \quad (1.30)$$

Three-phase power converters are superior to their single-phase counterpart, as they do not generate third harmonic currents. However, they still generate harmonic currents at their characteristic frequencies. Here, the characteristic frequencies are the integral multiples of the fundamental frequency, present in the Fourier series expansion of the converter's AC side current.

Figure 1.3 shows a three-phase six-pulse power converter, which is most commonly used in the 2–1,000 kVA range, with voltage rating ranging from 220 V to 13.8 kV. This power converter can be operated as a line-commutated unit or a forced-commutated unit, where diodes can be silicon-controlled rectifiers (SCR) or gate turn-off (GTO) thyristors. In forced-commutated units, control signals are used to turn on the switches (SCR or GTO), while the supply voltage determines the turn-off point. For a line-commutated unit, the phase A current, i_a, follows phase A voltage, v_{an}, when phase A voltage is the smallest of the three-phase voltages. The other phase currents are generated similarly. All three-phase currents and voltages for thc three-phase six-pulse converter are shown in Fig. 1.4.

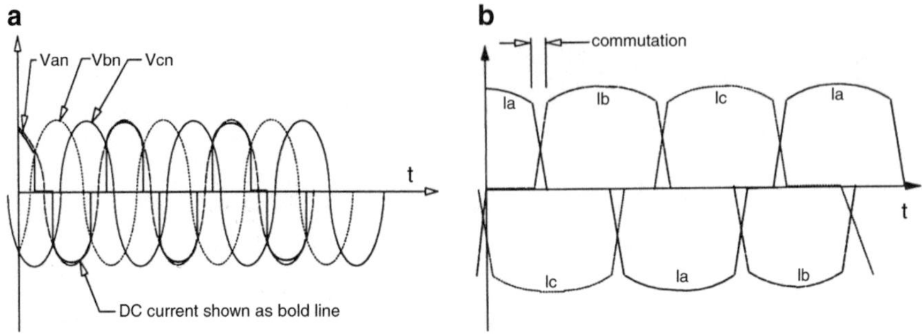

Fig. 1.4 (**a**) Voltages in three-phase six-pulse power converter. (**b**) Corresponding three-phase currents [1]

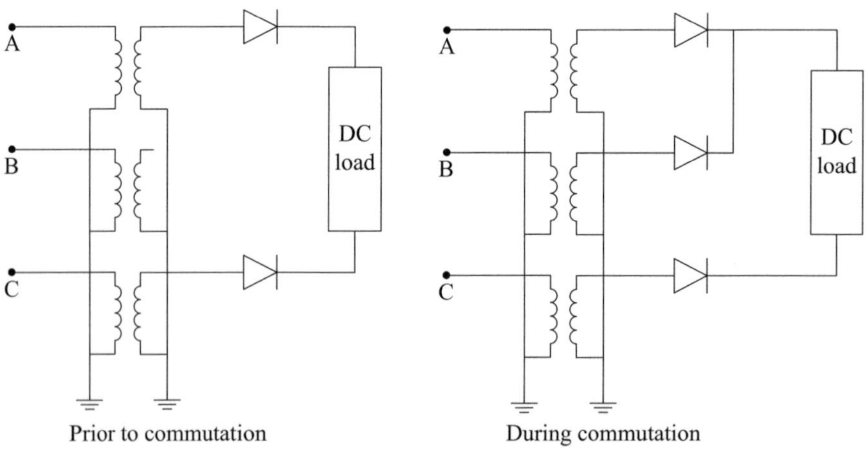

Fig. 1.5 Equivalent circuits of the three-phase six-pulse power converter, before and during commutation [1]

The operating principle for the three-phase six-pulse converter is as follows. For a time period of $0 < \theta < \frac{\pi}{3}$, the current i_a flows in the positive leg of the DC circuit. At $\theta = \frac{\pi}{3}$, the current path changes from phase A to phase B. This is referred to as commutation (see Fig. 1.5). Commutation is generally not instantaneous and is typically in the range of 0.05–1.4 ms. The practical effects of commutation relate to the harmonic impacts of the commutator. The details on commutation effects can be found in [1].

Next, the characteristics of the harmonics produced by the rectifier are discussed. The waveshapes of the phase currents are analyzed and Fourier series expansion is performed. The phase currents $(i_a(t), i_b(t), i_c(t))$, shown in Fig. 1.4, appear on both the primary and secondary sides of the transformers. This is because the DC components of the currents are absent. High inductance of the DC circuit causes I_{dc} to be fixed. The Fourier series of the phase A current is given by

$$i_a(t) = \frac{2\sqrt{3}}{\pi}\cos t + \frac{2\sqrt{3}}{\pi}\sum_{n=1}^{\infty}\left(\frac{(-1)^n}{6n-1}\cos(6n-1)t + \frac{(-1)^n}{6n+1}\cos(6n+1)t\right). \tag{1.31}$$

Clearly, the ideal six-pulse three-phase converter with $L_{dc} = \infty$ will induce harmonic currents on the AC side of the order of only, $6n \pm 1, n = 0, 1, \ldots$, termed characteristic harmonics.

1.3.2 Rotating AC Machines

Generally the pole faces of rotating electric machines are designed such that the low order harmonics in the supply will cancel. Depending upon the machine type, the harmonic cancelation can be done for up to the 11th-order harmonics. Furthermore, due to symmetry, AC machines produce no even order harmonics.

1.3.3 Fluorescent Lightning

Fluorescent lamps are very efficient sources of lighting because unlike incandescent lamps they do not dissipate much energy as heat. Therefore, the electrical energy supplied to the lamp is converted very efficiently to light. The fluorescent lamp works on the principle of gas ionization. Prior to when the gas in the fluorescent tube ionizes, the tube is an open circuit. After the gas ionization, the tube voltage drops dramatically. The exact V–I characteristics of the tube depends upon the length of the tube, the pressure, and type of gas. This is shown in Fig. 1.6. Figure 1.6 also shows that the current waveform *i*(*t*) is rich in third-order harmonics.

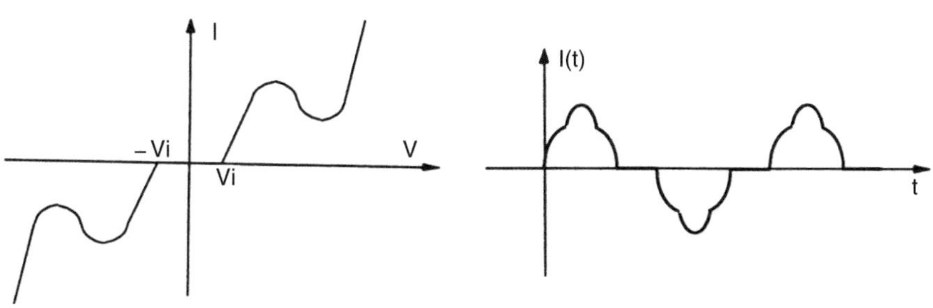

Fig. 1.6 V–I characteristics of a typical fluorescent lamp [1]

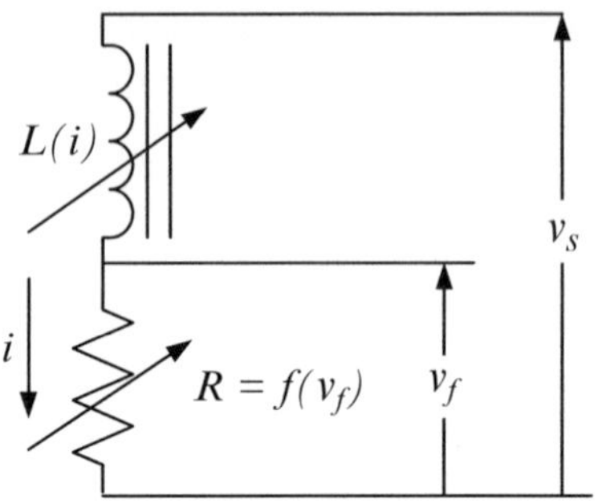

Fig. 1.7 Nonlinear equivalent circuit of a fluorescent lamp

Table 1.1 Current harmonics in a florescent lamp

Harmonic	Amplitude	Harmonic	Amplitude
1	1.00	6	0.01
2	0.04	7	0.05
3	0.20	8	0.00
4	0.01	9	0.06
5	0.10	10	0.00

Next, we will discuss the harmonic analysis of the current waveform produced by the fluorescent lamp. In the literature, three different approaches are used to calculate current in a fluorescent lamp [1]. In the first approach, a nonlinear resistance model for the tube is used to approximate the current waveform. Using a nonlinear resistance model, the current $i(t)$ can be represented as a nonlinear function of the voltage v_f, as shown in (1.32).

$$i = f(v_f) = A_0 + A_1 v_f + A_2 v_f^2 + \cdots . \tag{1.32}$$

Another method calculates the current by applying Kirchhoff's voltage law to the equivalent circuit model for the fluorescent lamp, shown in Fig. 1.7. On applying Kirchhoff's voltage law to this circuit, we obtain

$$\begin{aligned} v_s(t) &= \frac{d}{dt}(L(i(t))i(t)) + v_f(t) \\ v_f(t) &= v_s(t) - \frac{d}{dt}(Lf(v_f(t))) \end{aligned} \tag{1.33}$$

The current is obtained by iteratively solving differential equations given by (1.33). However, this method has convergence issues.

The third method calculates the current by simply using the measured historical data for the current harmonics produced by a fluorescent lamp. However, this approach is data intensive and the results are subject to question. Typical current harmonic amplitudes in a single-phase fluorescent lamp circuit are shown in Table 1.1.

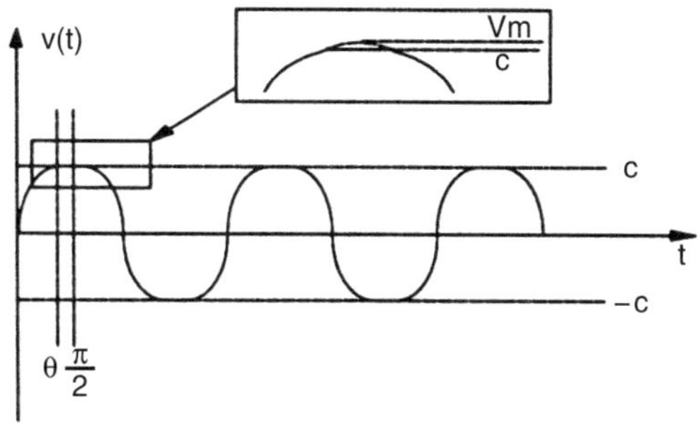

Fig. 1.8 Saturated transformer terminal voltage [1]

1.3.4 Overexcited Transformer

Overexcited transformers result in harmonic voltages. In addition, the harmonic impact of the overexcited transformer increases with the increase in the degree of overexcitation. Overexcitation of the transformer results in a clipped sine waveform as shown in Fig. 1.8. To understand the harmonic impact, the Fourier series of the clipped voltage waveform (with clipping starting at angle θ) is discussed. Note that the Fourier series only consists of sine terms, with coefficient of the nth harmonic given by b_n.

$$\begin{aligned} b_n = &\frac{V_m}{\pi}\left(\frac{\sin((n-1)\theta) - \sin((n-1)(\pi-\theta))}{n-1}\right. \\ &+\left.\frac{\sin((n+1)(\pi-\theta)) + \sin((n+1)\theta)}{n+1}\right) \\ &+\frac{V_m}{\pi}\left(\frac{2(\sin(\theta)\cos(n\theta) - \sin(\theta)\cos((\pi-\theta)n))}{n}\right), n \neq 1. \end{aligned} \tag{1.34}$$

The coefficient of the fundamental component is given by

$$b_1 = \frac{2V_m}{\pi}(\theta + \cos(\theta)\sin(\theta)) \tag{1.35}$$

1.3.5 Transformer Magnetization Current

Non-sinusoidal magnetizing current is also a source of harmonics in transformers. Transformer magnetizing current depends upon the third harmonic component. At full load, the amplitude of magnetizing current is below 2 % of the rated current. At low load or light load, the harmonic distortions in the magnetizing current can be well above 30 %. Figure 1.9 shows a typical characteristic of the harmonic content of the magnetizing current vs. the magnetic flux density (B) for different harmonic orders. From the figure, the harmonic content of the magnetizing current is a

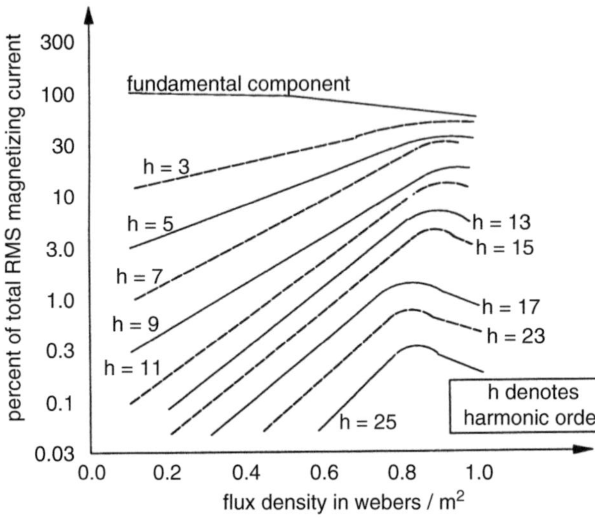

Fig. 1.9 Harmonic content of transformer magnetizing current [1]

nonlinear function of the magnetic flux density (*B*), and the harmonic order (*h*). The harmonic distortion due to a particular harmonic order (*h*) depends upon the transformer design. Furthermore, it can be observed from Fig. 1.9 that the transformers with their nominal design conditions located closer to the knee of the *B-H* curve will record a higher harmonic distortion. In other words, transformers nominally operating closer to the saturated region may record higher harmonic distortions.

1.3.6 Adjustable-Speed Drives

Adjustable Speed Drives (ASD) are power electronics based controllers used to control the speed of a rotating machines. These drives have been deployed in micromachines, household applications, and in commercial and industrial applications, thus having a control range from microwatts up to megawatts. In many applications ASDs help in reducing active power, thus decreasing the peak load demand and increasing the energy efficiency. ASDs also increase the efficiency of the machine and thereby improve the machine's coefficient of performance (COP). For example, in a residential ASD heat pump, the three-phase variable speed controller is used to adjust the machine's operating point while aiming to maximize the COP. Since maximum COP is targeted by the ASD's control mechanism, the energy efficiency of the heat pump increases significantly. Of course, the increased energy efficiency comes with the cost of increased harmonic distortion in the distribution system. The THD in the demand current can increase to as high as 70 %, thus severely affecting the distribution system power quality.

Table 1.2 Power quality impacts of commonly used adjustable-speed drives (ASDs) [1]

ASD type		AC line current	Harmonics	PQ impacts
AC variable voltage	Variable voltage constant frequency	Sine wave at power frequency modulated by square wave	All multiples of power frequency	High
AC variable frequency	Series and parallel chopper	(1) Chopped sine wave or (2) 6- or 12-pulse rectifier	(1) All multiples of fundamental frequency plus multiples of asynchronous frequency, or (2) $pn \pm 1$	(1) High, or (2) moderate
	Cyclo converter	Asynchronously chopped sine wave	Multiples of fundamental and cycloconverter frequency	High
	Rectifier and pulse modulator	6- or 12-pulse rectifier	$pn \pm 1$	Moderate
DC	DC machine drive	6- or 12-pulse rectifier	$pn \pm 1$	Moderate
AC slip recovery	Rotating Kramer and Scherbius	Sine wave	Fundamental only	None
	Static Kramer	6- or 12-pulse rectifier in rotor	$pn \pm 1$	Moderate
	Static Scherbius	Asynchronously chopped sine wave in rotor	Multiples of fundamental and cycloconverter frequency	Moderate to high

Several ASDs, controlling the speed of DC and AC machines and their impacts on the distribution power quality, are discussed in [1]. The book includes a detailed account on the circuit, control mechanism, and applications of various ASDs. The drives discussed in the book include DC machine drives, AC variable frequency drives, AC variable voltage drives, and AC slip recovery drives. Since the control as well as the operational mechanism is based on power electronic circuits, these drives inject significant harmonic currents into the system, resulting in power quality issues. A summary of the power quality issues caused by the most commonly used ASDs is presented in Table 1.2.

1.4 Networks and Component Modeling

This section covers the modeling of power system components in the presence of harmonics. Electric circuits are primarily modeled using two matrix formulations: bus admittance matrix (Y_{bus}) and bus impedance matrix (Z_{bus}). The general form of bus admittance matrix (Y_{bus}) is calculated by applying Kirchhoff's current law at each node (bus). The expression after applying Kirchhoff's current law at a node i in a general distribution network is given by (1.36).

$$\sum_j \overline{y_{ij}}\left(v_i - v_j\right) - i_i = 0, \tag{1.36}$$

where $\overline{y_{ij}}$ is the line impedance between nodes i and j, v_i and v_j are voltages at nodes i and j, respectively, and summation is carried out over all buses (j) present in the network.

Let the individual bus to ground voltage and currents injected at each node in the network be given by

$$V_{\text{bus}} = \begin{pmatrix} v_1 \\ v_2 \\ \vdots \\ v_n \end{pmatrix} \quad I_{\text{bus}} = \begin{pmatrix} i_1 \\ i_2 \\ \vdots \\ i_n \end{pmatrix}, \tag{1.37}$$

where n is the total number of buses.

Then using Kirchhoff's current law it is possible to write

$$I_{\text{bus}} = Y_{\text{bus}} V_{\text{bus}}. \tag{1.38}$$

Y_{bus} is constructed as follows:

- Y_{ii} is the sum of primitive line admittances connected to bus i.
- Y_{ij} is the negative of the primitive line admittance between buses i and j.
- the ground bus is used as reference bus and is not included in Y_{bus}.

The bus impedance matrix is simply the inverse of the bus admittance matrix and is given by (1.39).

$$Z_{\text{bus}} = Y_{\text{bus}}^{-1}. \tag{1.39}$$

Here, the diagonal elements of Z_{bus} are the driving point impedances at the system buses, and the off diagonal elements are the transfer impedances between two buses.

1.4.1 Transmission and Distribution System Under Non-sinusoidal Conditions

Many applications may require the bus impedance (Z_{bus}) and admittance (Y_{bus}) matrices to be evaluated at different frequencies. The Y_{bus} can be simply obtained

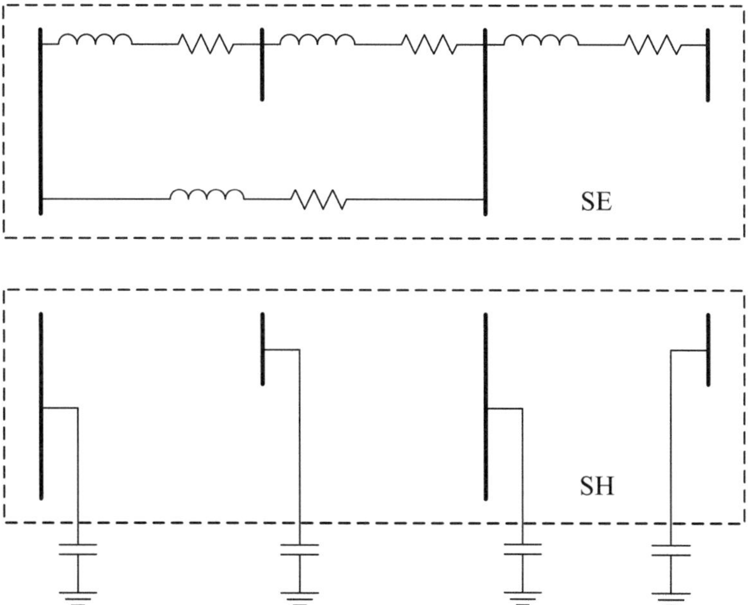

Fig. 1.10 Series and shunt partial networks

by applying the Y_{bus} building algorithm repeatedly after scaling the component data based on the harmonic frequencies as necessary. For example, transmission lines are modeled with constant R and constant L, transformer leakage reactance is scaled linearly with the harmonic frequency assuming constant leakage inductance, and core losses in the transformer are modeled as $\frac{R_c}{h}$ for harmonic order h.

An alternative formulation using the parallel impedance formula is also presented here. In this formulation, the distribution network is partitioned into two partial networks: one is a series network (SE) comprising of only series components, and the other is a shunt network (SH) with only shunt components. Both partial networks are shown in Fig. 1.10. Now, SE and SH partial networks are placed in parallel by joining all the corresponding buses, as shown in Fig. 1.11.

Clearly,

$$V = \begin{pmatrix} v_1 \\ v_2 \\ \vdots \\ v_n \end{pmatrix} = V_{\text{SE}} = V_{\text{SH}}, \tag{1.40}$$

where V_{SE} and V_{SH} are the bus voltage vectors for SE and SH networks, respectively.

Let the total current entering the two partial networks be the sum of the SE and SH current vectors.

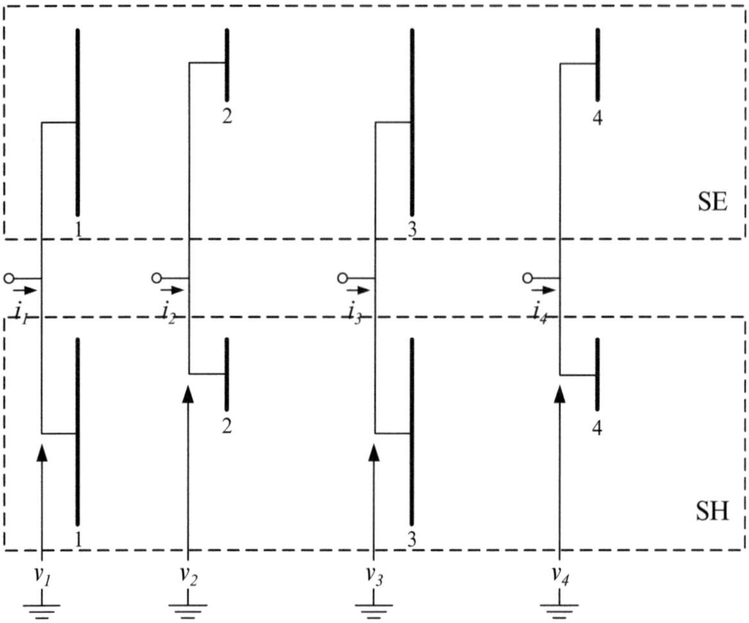

Fig. 1.11 The series and shunt partial networks connected in parallel

$$I_{eq} = \begin{pmatrix} i_1 \\ i_2 \\ \vdots \\ i_n \end{pmatrix} = I_{SE} + I_{SH}. \tag{1.41}$$

Since,

$$\begin{aligned} V_{SE} &= Z_{SE} I_{SE} \\ V_{SH} &= Z_{SH} I_{SH} \end{aligned}, \tag{1.42}$$

where Z_{SE} and Z_{SH} are impedance matrices for the SE and SH partial networks, the total current is given by

$$I_{eq} = \left(Z_{SE}^{-1} + Z_{SH}^{-1}\right) V. \tag{1.43}$$

Therefore, the equivalent impedance matrix Z_{eq} is given by

$$Z_{eq} = \left(Z_{SE}^{-1} + Z_{SH}^{-1}\right)^{-1}. \tag{1.44}$$

1.4.2 Shunt Capacitor

A shunt capacitor is modeled as a reactance connected between the bus and the ground and is included while building the system Y_{bus} and Z_{bus} matrices. It is very important to include the shunt capacitance at higher frequencies. At higher frequencies, capacitive elements become low impedance paths and may induce resonance in the circuit response. The term resonance is defined as an operating condition such that the magnitude of the impedance of the circuit passes through an extremum. For example, a simple LC circuit has impedance:

$$Z(\omega) = j\omega L + \frac{1}{j\omega C}. \tag{1.45}$$

The magnitude of this function, shown in (1.45) has a minimum at $\omega_{\min}$ (resonant frequency), given by (1.46).

$$\omega_{\min} = \frac{1}{\sqrt{LC}}. \tag{1.46}$$

Note that the impedance also achieves an extremum as $\omega_{\max} \rightarrow (0, \infty)$.

The circuit resonance are of two types: series and parallel. Series resonance occurs when impedance function passes through a minimum, and parallel resonance occurs when impedance function passes through a maximum. The methods to calculate the resonance conditions are summarized as follows.

The search of resonance condition amounts to the search for the extrema of the impedance function $Z(\omega)$. One method is to calculate the circuit impedance for a range of frequencies and scan for a minimum and/or maximum values for $|Z(\omega)|$. Another method is to examine the phase of the impedance function. Resonance may occur for the system frequency when the phase angle passes through zero.

1.4.3 Transformer

The commonly used T-equivalent circuit of a single-phase transformer is shown in Fig. 1.12. The transformer circuit parameters are calculated using open circuit and short circuit tests. Both open circuit and short circuit tests are performed at the

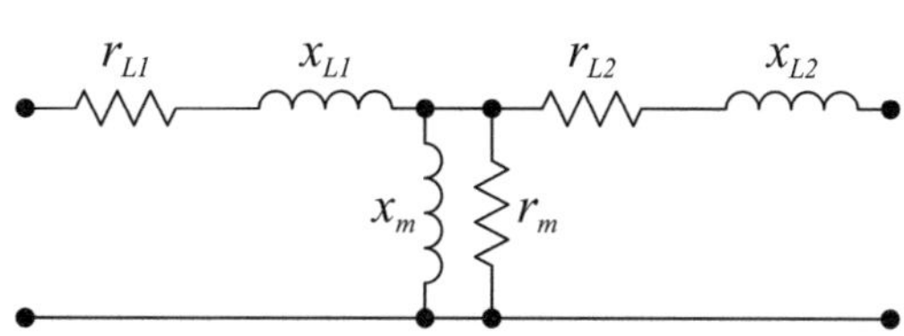

Fig. 1.12 T-equivalent circuit of a transformer

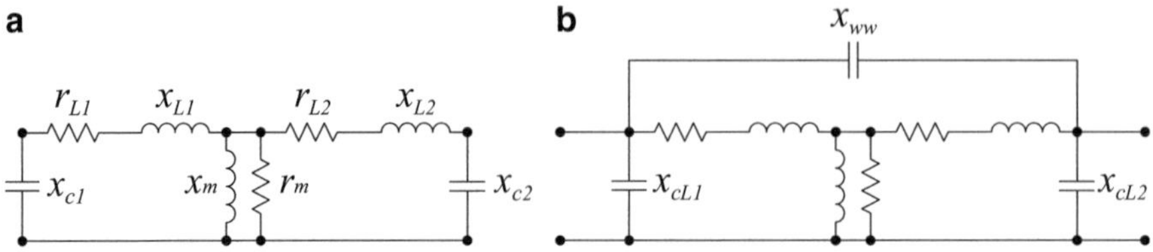

Fig. 1.13 (**a**) Second order model of a transformer. (**b**) Modified second order model

power frequency and the equivalent circuit parameters are expressed at the power frequency. Note that, because most power transformers are designed to operate only at the power frequency, their frequency response drops with both an increase or decrease in the frequency. For example, the leakage reactance of the transformer increases with an increase in frequency. Similarly, transformer core losses increase with an increase in frequency. Also, the transformer secondary voltage drops as the frequency decreases.

In order to include and expand upon the frequency response characteristics of the transformer, it is required to increase the model complexity. A second order model is shown in Fig. 1.13a. The capacitive reactance in the model represents the lumped model of the turn-to-turn capacitance, stray lead capacitance, and the capacitance from the winding to the case. Because the capacitance provides a low impedance path to high frequencies, this model simulates the attenuation of high frequencies by the transformer capacitance. Furthermore, the second order transformer model can be modified to include winding-to-winding capacitance (see Fig. 1.13b).

The next level of sophistication for a transformer model is obtained by modeling the transformer inductance and capacitance as distributed elements. This model is much more complex than the lumped parameter model discussed before and is simulated only for rather specific applications. One such model is shown in Fig. 1.14.

In this model:

- $r_{L1}(k)$ and $r_{L2}(k)$—leakage reactance of kth turn section of primary and secondary windings, respectively.
- $M(k)$—mutual impedance between primary and secondary circuit at section k.
- $x_{c1}(k)$ and $x_{c2}(k)$—turn-to-turn capacitance for primary and secondary windings, respectively.
- $x_{cw}(k)$—winding-to-winding capacitive reactance at section k.
- x_{cL1}—lead capacitance, the brushing capacitance and lumped winding to case capacitance in the primary.

Note that the parameter k is taken over n sections of windings. Also, as n becomes large, this model becomes a distributed parameter model.

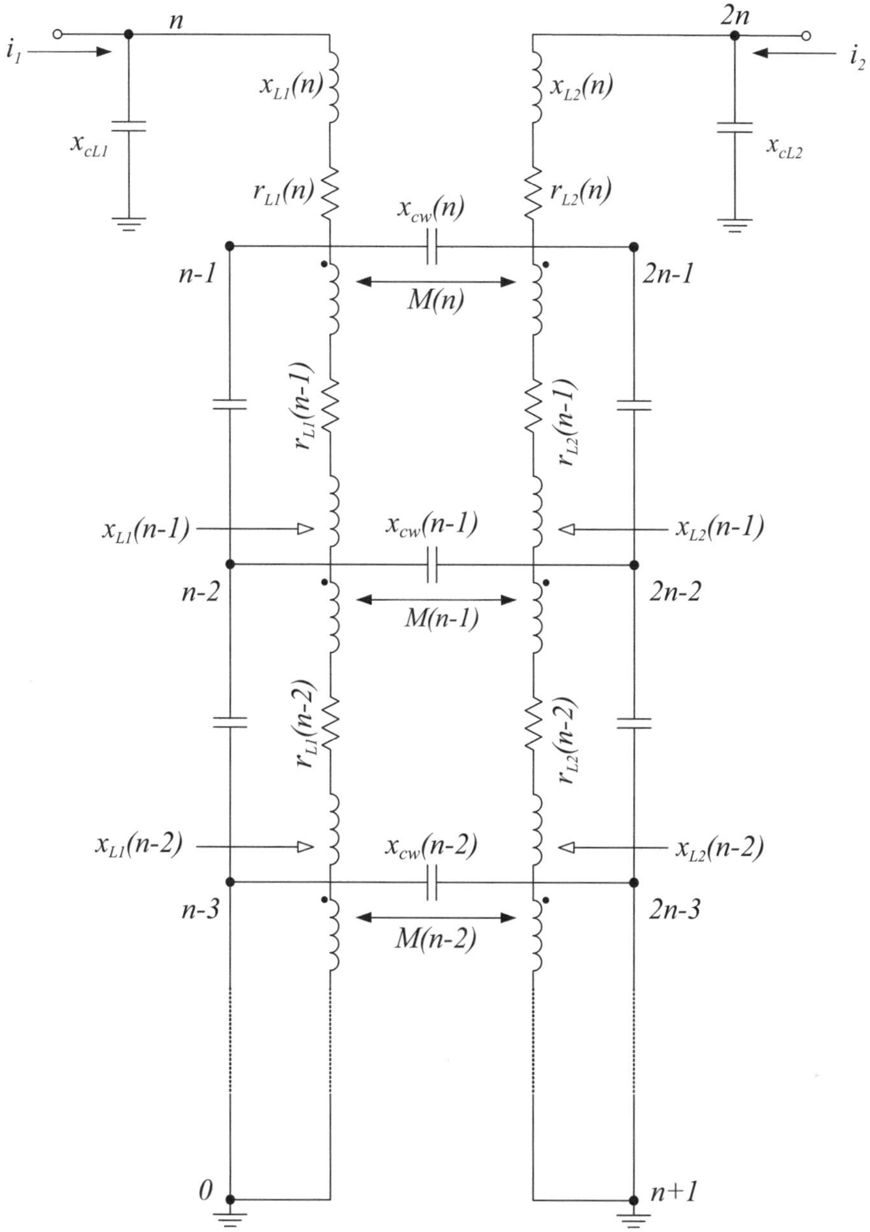

Fig. 1.14 Distributed parameter model of a transformer

1.4.4 Electric Machine

In this section, the electric machine models are discussed when operating under non-sinusoidal conditions. One of the primary power quality concerns due to electric machines is a high startup current resulting from the inrush phenomenon. Inrush currents are induced due to the energization of an inductive or capacitive circuit from an AC bus. Therefore, inrush currents have the characteristics similar to that of an RL or RC circuit. The inrush current, which is the natural component of the transient startup current, typically decays in the range of 1–100 s. The high current in a rotating machine may also be the result of the machine starting current, which is generally higher than the inrush current. Note that the high starting current in a rotating machine originates because of the absence of a "speed voltage" or back EMF while starting the machine.

Next, we will discuss the characteristics of induction motors under non-sinusoidal conditions. Under sinusoidal conditions, the induction machine is modeled as a simple reactance for positive sequence voltages, x^+, and negative sequence voltages, x^-. For zero sequence voltage, the machine will appear as an open circuit if connected in ungrounded wye or in delta. If the machine is connected in a grounded wye, the zero sequence impedance, z_0, will be very high and will be characterized by a magnetizing and heat loss phenomenon.

For non-sinusoidal voltages, an induction motor reacts to the harmonic voltages according to the sequence components. Let s denote the slip of the induction machine, given by

$$s = \frac{\omega_0 - \omega_r}{\omega_0}, \tag{1.47}$$

where ω_r is the rotor mechanical speed, and ω_0 is the stator MMF wave speed.

The rotor current frequency, ω_{lr}, due to a bus voltage with harmonic order h producing positive sequence current, as shown in Fig. 1.15a, is given by (1.48).

$$\omega_{lr} = (h - 1 + s)\omega_0 \tag{1.48}$$

while, for a bus voltage producing negative sequence current (Fig. 1.15b), induces rotor current frequency given by (1.49).

$$\omega_{lr} = (h + 1 - s)\omega_0. \tag{1.49}$$

Note that the rotor appears to the stator as an impedance. At harmonic frequencies, the magnitude of the rotor impedance for positive and negative sequence bus voltages is given by

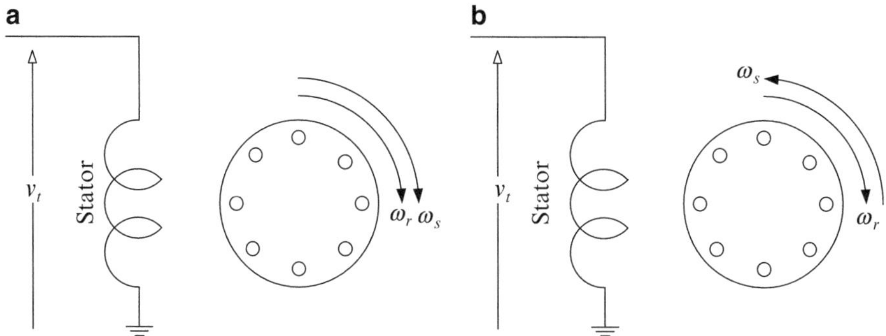

Fig. 1.15 (**a**) Induction motor response to positive sequence bus voltage. (**b**) Induction motor response to negative sequence bus voltage

$$x = (h \pm 1 - s)x^{-}. \tag{1.50}$$

In the above expression (1.50), the negative sequence reactance is calculated at fundamental frequency and the upper sign corresponds to positive sequence while the lower sign corresponds to the negative sequence.

Following the above discussion, the simplified model of an inductance motor for harmonic components in the supply voltage is a simple reactance calculated using (1.50). This model does not include active power losses and B-H losses.

1.5 Methods Used for Harmonic Analysis

Once we have discussed the modeling of power system components under non-sinusoidal conditions, the next step is to describe the methods used for harmonic analysis. The harmonic analysis is done using harmonic power flow. The objective of harmonic power flow calculations is to determine bus harmonic voltages in a given power system while harmonic sources are present in the network. The calculated harmonic voltages are used to determine the network's voltage and current THD. Furthermore, corresponding voltage and current waveforms can also be obtained by superimposing different harmonic components. Three popular methods used for harmonic analysis are described in this section:

- Current Injection Method
- Harmonic Power Flow using the coordinate method, and
- Harmonic Power Flow using the Newton–Raphson method

The current injection method is a non-iterative method, formulated using a current injection model for the harmonic sources. Injected harmonic currents are calculated using the bus voltages obtained by executing the power flow algorithm under sinusoidal conditions. The method is reasonably accurate under low bus voltage THD and low injected current THD. The accuracy of the method decreases on increasing harmonic distortions.

The other two methods are based on conventional power flow studies. A power flow study is an algorithm used to analyze the network for a given load, generation condition, and given circuit impedances. These data are used together to calculate bus voltages of the network and power flows in the individual distribution circuit lines. The power flow algorithms and methods to solve them are summarized in this section.

To execute a power flow algorithm, the following information is initially specified:

- Active and reactive power are specified at load buses (also called PQ buses).
- Active power and bus voltages are specified at generation buses (also called PV buses).
- The bus impedance or bus admittance matrix for the network is specified.
- At least one bus of the system must have unspecified power. This bus is referred to as the swing bus. At the swing bus, the bus voltage magnitude is specified. Also the swing bus is chosen as the reference phasor with known voltage phase equal to zero.

The parameters to be calculated using the power flow algorithm are as follows:

1. Bus voltages.
2. Line active and reactive power flows.

There are two methods to solve the power flow problem. One is the coordinate method and the other is using a mismatch formulation. The coordinate method is based on the definition of the terms PQ and PV bus, and the network equations; it is solved using the Gauss–Seidel method. The mismatch formulation is based on the solution of a set of nonlinear equations and is solved using the Newton–Raphson method. In the following section, both the coordinate method and the mismatch method are formulated and solved for the conventional power flow study executed at the rated power frequency. Both methods are extended in the case of harmonics in the subsequent sections.

Conventional Power Flow Using the Coordinate Method

The power flow problem formulation using the coordinate formulation is discussed here. The given network information is as follows:

At PQ Bus

$$v_i i_i^* = P_i + jQ_i. \tag{1.51}$$

At PV Bus

$$\begin{aligned} |\ v_i\ | &= \text{specified} \\ \text{Re}(v_i i_i^*) &= P_i. \end{aligned} \tag{1.52}$$

Table 1.3 Types of buses and number of unknowns in a conventional power flow formulation

Real equations		Real unknowns	
PV bus	$N_{pv} = 2n_{pv}$	V_{bus}	$2N$
PQ bus	$N_{pq} = 2n_{pq}$	I_{bus}	$2N$
Swing bus	2		
Network	$2N$		
Total	$N_{pv} + N_{pq} + 2 + 2N = 4N$	Total	$4N$

At swing Bus

$$v_{swing} = \mid v_{known} \mid \angle 0°. \tag{1.53}$$

The Network

$$V_{bus} = Z_{bus} I_{bus}. \tag{1.54}$$

Equations (1.51)–(1.54) constitute $4N$ real equations in $4N$ real unknowns. Here N is equal to the number of system buses. All equations and unknowns for the power flow study are summarized in Table 1.3.

The coordinate formulation for the power flow study is solved using the Gauss–Seidel method. The Gauss–Seidel method is an iterative procedure that is performed as follows:

1. First, the circuit unknowns are initialized. The initial value of bus voltages is assumed unity with zero phase angle. The line currents are initialized and are equal to the complex conjugate of the specified bus power.
2. The PQ equations, PV equations, and network expressions are solved for bus voltages. The network expressions are used to sequentially update the bus voltages and currents. The first expression updates the first voltage and the next expression uses the first updated voltage to update the first current and second voltage, etc.
3. In this way, all the bus voltages and currents are updated once in the first iteration. In each update for voltage and current, the algorithm uses the most recently updated value of all other variables.
4. The iterative process is repeated after updating all the variables.
5. The process is stopped when the change in variables (bus voltages) after an update decreases below a specified tolerance value.

The Gauss–Seidel solution requires several iterations to solve the problem; therefore, it is computationally expensive. Furthermore, the convergence property of this method is not strong; this also applies to harmonic power flow using the coordinate method.

Conventional Power Flow Using the Newton–Raphson Method

In this section, the power flow studies using the Newton–Raphson method are discussed. First, the conventional power flow algorithm is discussed. The power flow problem is written as a mismatch formulation and solved using the Newton–Raphson method. The mismatch formulation aims to find the solution to a set of simultaneous equations of the form:

$$F(X) = 0, \tag{1.55}$$

where both F and X are vector valued.

In the case of the conventional power flow problem, vector F is the mismatch between active and reactive power for each bus, except the swing bus. At a solution step, the mismatch is calculated by adding the active and reactive power entering a bus. For example, at a bus i, the complex power may be entering the bus due to generation and loads at that bus, and also from all other network buses that are connected to bus i. Let j denote a bus connected to the bus i and S_i be the complex power at bus i due to load and generation at bus i. Then, the complex power mismatch is given as

$$\Delta S_i = S_i + \sum_j (v_j - v_i)^* \overline{y_{ij}^*} v_i, \tag{1.56}$$

where $\overline{y_{ij}}$ is primitive line admittance between buses i and j and sum is carried over all buses $j \forall j \neq i$.

The active and reactive power mismatch at bus i are

$$\begin{aligned} \Delta P_i &= P_i - |Y_{ii}||v_i||v_i| \cos(-\theta_{ii}) - \sum_{j \neq i} |Y_{ij}||v_i||v_j| \cos(-\theta_{ij} - \delta_j + \delta_i) \\ \Delta Q_i &= Q_i - |Y_{ii}||v_i||v_i| \sin(-\theta_{ii}) - \sum_{j \neq i} |Y_{ij}||v_i||v_j| \sin(-\theta_{ij} - \delta_j + \delta_i), \end{aligned} \tag{1.57}$$

where the bus admittance matrix is written in polar form, and the ij^{th} element of the bus admittance matrix is $Y_{ij}|\angle\theta_{ij}$.

Also, bus voltages are written in polar form, with bus i and bus j voltage equal to $v_i\angle\delta_i$ and $v_j\angle\delta_j$, respectively. The terms P_i and Q_i are the specified active and reactive power at bus i.

The mismatch equation is then formulated as follows:

$$\begin{pmatrix} \Delta P(\delta, V) \\ \Delta Q(\delta, V) \end{pmatrix} = 0, \tag{1.58}$$

where $\begin{pmatrix} \Delta P(\delta, V) \\ \Delta Q(\delta, V) \end{pmatrix}$ is the array of power mismatch for $N - 1$ buses (excluding the swing bus), and δ and V are the $N - 1$ dimensional arrays of bus voltages and phase angles for the $N - 1$ buses.

The method to solve the power flow mismatch equation using the Newton–Raphson method is summarized in the following section.

Let

$$F = \begin{pmatrix} \Delta P(\delta, V) \\ \Delta Q(\delta, V) \end{pmatrix} = 0. \tag{1.59}$$

Next, (1.55) is expanded in a Taylor series about a vector, $X = \begin{pmatrix} \delta \\ V \end{pmatrix}$, representing the initial guess.

$$F(X) = F(X_0) + \frac{\partial F}{\partial X}\bigg|_{X_0} (X - X_0) + \text{higher order terms}. \tag{1.60}$$

Solving (1.60) for X while $F(X) = 0$, gives,

$$X_1 = X_0 - \left(\frac{\partial F}{\partial X}\bigg|_{X_0} \right)^{-1} F(X_0), \tag{1.61}$$

where X_1 denotes approximate value of the vector X evaluated at first iteration.

Note that an approximate value of the variable is obtained by N–R method, as the Taylor series is truncated at the first derivative. Equation (1.63) is called the Newton–Raphson update formula, which is used to solve for $F(X) = 0$.

In the update formula, the partial derivative term $\left(\frac{\partial F}{\partial X}\right)$ is called the Jacobian matrix. The general Newton–Raphson update equation from k to $k + 1$ iteration is given by

$$X_{k+1} = X_k - \left(\frac{\partial F}{\partial X}\bigg|_{X_k} \right)^{-1} F(X_k). \tag{1.62}$$

The elements of the Jacobian matrix are obtained by partial differentiation of (1.57) with respect to vector δ and V. Equations (1.63) and (1.64) give the formulas to calculate the Jacobian matrix.

$$\begin{aligned}
\frac{\partial(\Delta P_i)}{\partial \delta_i} &= \sum_{j \neq i} |Y_{ij}||v_i||v_j| \sin\left(-\theta_{ij} - \delta_j + \delta_i\right) \\
\frac{\partial(\Delta P_i)}{\partial \delta_k} &= -|Y_{ik}||v_k||v_i| \sin\left(-\theta_{ik} - \delta_k + \delta_i\right), i \neq k \\
\frac{\partial(\Delta P_i)}{\partial |v_i|} &= -2|Y_{ii}||v_i| \cos\left(-\theta_{ii}\right) - \sum_{j \neq i} |Y_{ij}||v_j| \cos\left(-\theta_{ij} - \delta_j + \delta_i\right) \\
\frac{\partial(\Delta P_i)}{\partial |v_k|} &= -|Y_{ik}||v_i| \cos\left(-\theta_{ik} - \delta_k + \delta_i\right), i \neq k,
\end{aligned} \tag{1.63}$$

$$\begin{aligned}
\frac{\partial(\Delta Q_i)}{\partial \delta_i} &= -\sum_{j \neq i} |Y_{ij}||v_i||v_j| \cos\left(-\theta_{ij} - \delta_j + \delta_i\right) \\
\frac{\partial(\Delta Q_i)}{\partial \delta_k} &= |Y_{ik}||v_k||v_i| \cos\left(-\theta_{ik} - \delta_k + \delta_i\right), i \neq k \\
\frac{\partial(\Delta Q_i)}{\partial |v_i|} &= -2|Y_{ii}||v_i| \sin\left(-\theta_{ii}\right) - \sum_{j \neq i} |Y_{ij}||v_j| \sin\left(-\theta_{ij} - \delta_j + \delta_i\right) \\
\frac{\partial(\Delta Q_i)}{\partial |v_k|} &= -|Y_{ik}||v_i| \sin\left(-\theta_{ik} - \delta_k + \delta_i\right), i \neq k.
\end{aligned} \tag{1.64}$$

The Jacobian matrix (1.88) is denoted by J, and is usually partitioned in four submatrices.

$$J = \begin{bmatrix} \frac{\partial(\Delta P)}{\partial \delta} & \frac{\partial(\Delta P)}{\partial |V|} \\ \frac{\partial(\Delta Q)}{\partial \delta} & \frac{\partial(\Delta Q)}{\partial |V|} \end{bmatrix}. \tag{1.65}$$

1.5.1 Harmonic Analysis Using the Current Injection Method

In the current injection method, the harmonic sources are modeled as current injection sources. Although the current injection model is only a first cut approximation, it is reasonably accurate at low levels of harmonic distortions and at nearly sinusoidal bus voltages. Mostly AC/DC converters, which are the primary source of harmonics in a power system, are modeled as current injection sources. The injection currents at different harmonic frequencies are assumed to be independent of each other. If harmonic current $i^{(h)}$ is injected into a bus, then by Ohm's law, the system voltage is given by the vector $V_{\text{bus}}^{(h)}$.

$$V_{\mathrm{bus}}^{(h)} = Z_{\mathrm{bus}}^{(h)} I_{\mathrm{bus}}^{(h)}, \tag{1.66}$$

where the vector $I_{\mathrm{bus}}^{(h)}$ is all zero except in the position that corresponds to the bus at which the harmonic current is injected.

The steps for applying the current injection model described in (1.66) are as follows:

- First, the conventional load flow analysis is performed for the given network.
- Next, using the conventional load flow study done in the previous step, a current source model for the converters connected to the load buses is developed. The bus voltages are used to calculate a set of harmonic currents $I_{\mathrm{bus}}^{(h)}$ at each load bus connected to the converter.
- Using the calculated harmonic currents and (1.66), the harmonic bus voltages are calculated for each harmonic.
- The harmonic voltages calculated in the previous step, for a given bus, are superimposed to obtain the bus voltage waveforms.

Instead of using the impedance matrix in (1.66), the harmonic voltages can also be calculated using the bus admittance matrix. Since the equation needs to be repeated for each frequency, it is generally not advisable to generate an impedance matrix each time. The harmonic voltage calculation algorithm using the bus admittance matrix is given in (1.67).

$$V_{\mathrm{bus}}^{(h)} = (Y_{\mathrm{bus}}^{(h)})^{-1} I_{\mathrm{bus}}^{(h)}. \tag{1.67}$$

The bus admittance matrix ($Y_{\mathrm{bus}}^{(h)}$) used in (1.67) is obtained using the bus admittance building algorithm, which is time and memory efficient. Furthermore, the matrix inversion can be avoided using triangular factors of Y_{bus}. The voltages are calculated using forward and backward substitution, as shown in (1.68).

$$\begin{aligned} I_{\mathrm{bus}}^{(h)} &= Y_{\mathrm{bus}}^{(h)} V_{\mathrm{bus}}^{(h)} \\ I_{\mathrm{bus}}^{(h)} &= (L)(U) V_{\mathrm{bus}}^{(h)} \\ I_{\mathrm{bus}}^{(h)} &= (L)(W) \\ W &= (U) V_{\mathrm{bus}}^{(h)} \end{aligned}. \tag{1.68}$$

In (1.68), L and U are the lower left and upper right triangular factors of the admittance matrix. The vector W is calculated by forward substitution, the vector $V_{\mathrm{bus}}^{(h)}$ using backward substitution.

The current injection method is very efficient in terms of speed and memory. Another advantage is that the method is non-iterative and always results in a solution. Since matrix inversion can be avoided, a full matrix does not need to be stored. Furthermore, at low bus voltage THD and low converter current THD, the

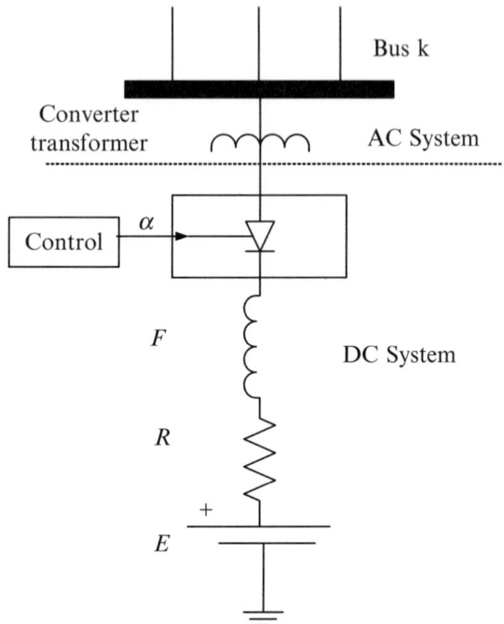

Fig. 1.16 Rectifier load at bus *k*

method is reasonably accurate. However, the current injection method is inaccurate at higher harmonic penetration, making the method ineffective at THD levels above 5 %. This is because in the current injection method, the injected current is calculated using a sinusoidal bus voltage waveform. At higher THD levels, voltage waveforms are no longer sinusoidal leading to inaccurate results.

1.5.2 Harmonic Power Flow Using the Coordinate Method

The harmonic power flow algorithm solved using the coordinate method is described in detail in this section. The conventional coordinate method based power flow algorithm requires modification when a nonlinear load is present in the system. It should be noted that harmonic load flow analysis requires considerable knowledge about each harmonic load because the harmonic signal level depends very strongly upon the exact load configuration. For example, the harmonic profile of a six-pulse converter will differ greatly from a 12-pulse converter.

In this section, the power flow algorithm is formulated for a single harmonic source. A rectifier load, specifically a six-pulse Graetz bridge, is assumed to be connected to the load bus *k*, as shown in Fig. 1.16. The controls for the converter's firing angle are modeled as follows:

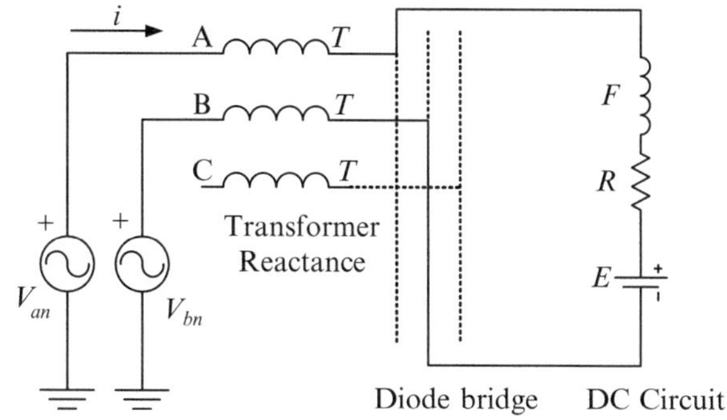

Fig. 1.17 Conduction in phases A and B of a three-phase six-pulse rectifier

$$\alpha = f(P_{\text{dc}}, |v_k|, |i_k|), \tag{1.69}$$

where P_{dc} is the output DC power.

As can be seen from Fig. 1.16, the DC side of the six-pulse converter is modeled as a fixed impedance ($R + j\omega F$) behind a DC voltage (E). Also, the transformer connecting the converter to the AC bus is modeled as a fixed inductance denoted as T.

During Conduction Period

Next, the circuit equations when the rectifier is operating during the conduction period are discussed. Figure 1.17 shows the equivalent circuit for the rectifier when phases A and B are conducting. The current waveform is determined by solving the differential equation for the conduction time period. The equation for the equivalent circuit shown in Fig. 1.17 is given below:

$$V_{an}(t) - V_{bn}(t) = (2T + F)\frac{di(t)}{dt} + i(t)R + E. \tag{1.70}$$

The solution of the expression is given by

$$i(t) = e^{-Rt/(F+2T)} i(t_0) + \frac{e^{-Rt/(F+2T)}}{(F + 2T)} * (V_{an}(t) - V_{bn}(t) - E), \tag{1.71}$$

where (*) denotes the convolution operator, and $i(t_0)$ is the DC current at the beginning of the time interval.

The expression given in (1.71) is initially solved using sinusoidal bus voltages. Later, the DC current solution under harmonic voltages is obtained by numerically solving the expression in (1.71). Numerical solutions can be obtained either by numerically solving the differential equation or by numerically evaluating the convolution expression for the DC current. Note that, for the harmonic solutions, the AC bus voltage is expressed as a Fourier series and the elements of the Fourier series are calculated for a specific time value.

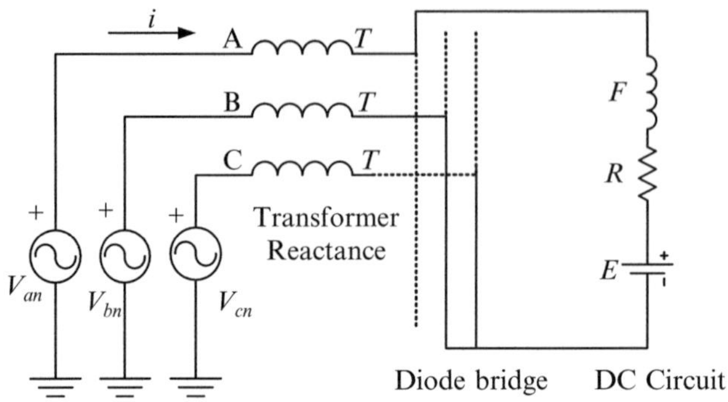

Fig. 1.18 Commutation in phases B and C of a three-phase six-pulse rectifier

During Commutation Period

In addition to the six conduction periods, the six-pulse converter also exhibits six commutation periods. An example equivalent circuit diagram of the converter for the commutation of phase B to phase C is shown in Fig. 1.18. During commutation, phase B and phase C are short-circuited. Again, using KVL, the first order differential equation for the converter in the commutation stage is derived.

$$-V_{an}(t)+\frac{1}{2}V_{cn}(t)+\frac{1}{2}V_{bn}(t)+i(t)R+E+\frac{di(t)}{dt}\left(F+\frac{3}{2}T\right)=0. \qquad (1.72)$$

And the solution is given as

$$i(t)=e^{\left(\frac{-Rt}{F+\frac{3}{2}T}\right)}i(t_0)+\frac{e^{\left(\frac{-Rt}{F+\frac{3}{2}T}\right)}}{\left(\frac{-Rt}{F+\frac{3}{2}T}\right)}*\left(V_{an}(t)-\frac{1}{2}V_{cn}(t)-\frac{1}{2}V_{bn}(t)-E\right). \qquad (1.73)$$

Power Flow Study

Note that similar to the conventional power flow, the load data for the rectifier is expressed in terms of the power consumption. Since the calculated phase current is a function of the firing angle, it is required to iteratively update the firing angle value so that the calculated solution fits the given rectifier load data. The iterative phase current calculation procedure is described as follows. First, the firing angle α is initialized and the phase currents are calculated for all six conduction and commutation intervals for the converter. Next, the average power over a cycle is calculated using the calculated current waveforms and the calculated power is compared against the specified load power. Based on the power mismatch between the calculated power and the specified power, firing angle α is updated. The current waveform is recalculated and the firing angle is updated until the mismatch between the calculated power and the specified power is below the specified tolerance.

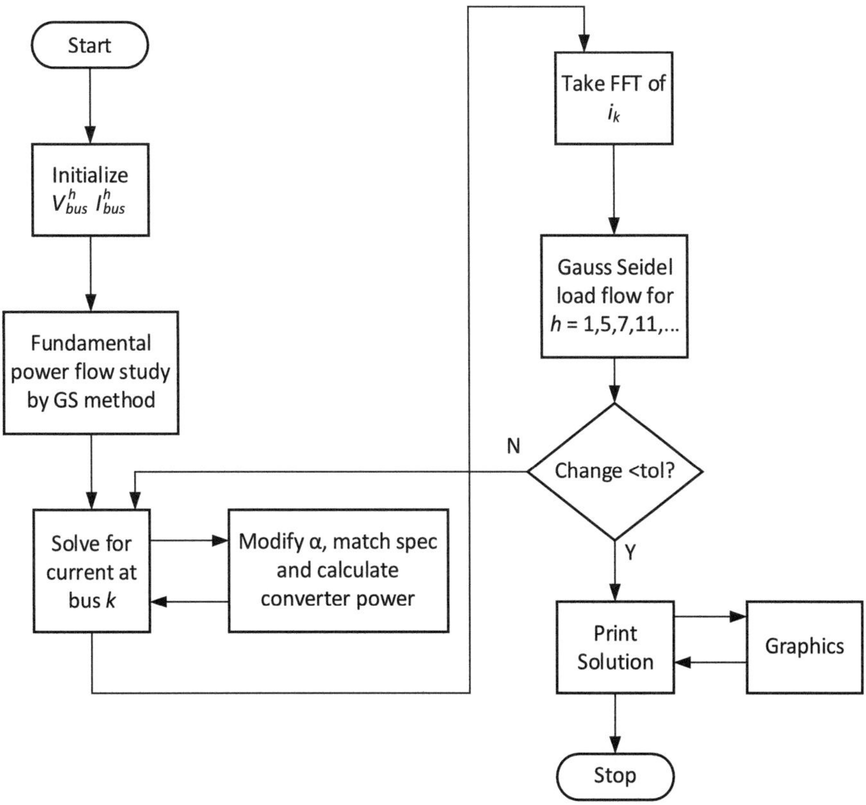

Fig. 1.19 Harmonic power flow study using coordinate algorithm

Alternatively, a closed form expression for the phase current waveform is developed for the conduction and commutation stages in (1.70)–(1.73). The closed form solution can be developed for all six conduction and commutation intervals and can be differentiated with respect to firing angle (α). This will result in a closed form expression for the firing angle update rule, as given by (1.74).

$$\alpha_{n+1} = \alpha_n + \frac{d\alpha}{dP}\Delta P. \tag{1.74}$$

Once the solution for the firing angle converges, the FFT of the phase current waveform is calculated. The FFT represents the harmonic content of the load current, for the load bus connected to the rectifier. This current is used to model the load for each harmonic frequency under consideration. Next, to solve the harmonic power flow, the Gauss–Seidel algorithm is executed for each harmonic frequency. Note that the system admittance matrix should be modified according to the harmonic frequency. The algorithm is described in detail in Fig. 1.19.

The convergence properties of the coordinate method based harmonic power flow are summarized here. Because of additional variables and complexity, a harmonic power flow generally takes more execution time than the conventional power flow. Overall, a Gauss–Seidel solution for the harmonic power flow takes more than $1.3H$ times longer than the corresponding solution to the conventional power flow (H is the number of harmonic frequencies under study). In summary, for the coordinate based harmonic power flow formulation:

- Each harmonic must be solved by the Gauss–Seidel algorithm.
- The operating point (firing angle α) must be solved iteratively.
- Initialization of the harmonic levels is difficult.
- If the initial value starts far off from the actual solution, the algorithm will diverge.

1.5.3 Harmonic Power Flow Using the Newton–Raphson Method

In this section, the Newton–Raphson based harmonic power flow algorithm is summarized. The Newton–Raphson method for power flow analysis has superior convergence properties over the coordinate method. Therefore, the Newton–Raphson method is extended for harmonic power flow analysis. The mismatch power expressions are augmented with the expressions representing mismatch in harmonic currents, therefore extending the fundamental frequency formulation to harmonic frequencies.

The discussion begins with characterizing the bus types used in the formulation of harmonic power flow algorithm. Next, the power mismatch equations under harmonic distortion will be derived and Jacobian matrix formulation under both the fundamental and harmonic scenario will be discussed. This discussion will entail a detailed explanation on state variables, number of equations, and harmonic equations. Next, the Newton–Raphson based algorithm to solve the formulated mismatch harmonic power flow problem will be discussed.

Types of Buses

Under harmonic power flow study, the power system buses are primarily categorized as either linear or nonlinear. When a sinusoidal voltage is applied, a linear power system bus results in a sinusoidal current. However, in the case of a nonlinear bus, the line currents are distorted even if the supply voltage is sinusoidal. The buses connected to the conventional loads exhibit linear characteristics, while the buses connected to the converters and other nonlinear loads (fluorescent lamps, ASD, etc.) are nonlinear. For the harmonic power flow formulation, four categories of power system buses are identified. The four bus types are discussed below:

1. Swing Bus—This is a single, linear, voltage regulated bus. Like conventional power flow studies, this bus is used as a reference bus.

2. Linear PQ Buses—These are the load buses where active and reactive power (P and Q) at fundamental frequency are specified. Note that, these buses are connected to the conventional loads and therefore are linear. There are n_{lpq} linear PQ buses.
3. Nonlinear PQ Buses—These buses are connected to unconventional loads, such as converters and other nonlinear devices. There are n_{nlpq} nonlinear PQ buses.
4. Linear PV Buses—These are the generation buses, where active power and voltage are specified. There are n_{lpv} linear PV buses.

Note that the system might contain nonlinear PV buses as well; for example, a voltage regulated bus is connected to an inverter. Here, the inverter parameters might be controlled for bus voltage regulation. This specific case can be modeled in a similar way as the linear PV buses and therefore is not specifically considered in the discussion.

Power Mismatch Formulation

As discussed before, in the case of conventional power flow for PQ buses, bus voltage magnitudes and angles are used as state variables. As for the PV buses, voltage magnitude is given, therefore, voltage angles need to be determined. For harmonic power flow, additional state variables are introduced to represent quantities corresponding to harmonic voltages and currents. The state variables corresponding to each bus type and power mismatch equations are described in this section.

Linear PQ Bus

At linear PQ buses, the conventional active and reactive power mismatch equations are given as follows:

$$\begin{aligned} \Delta P(|V_{\mathrm{bus}}|, \delta_{\mathrm{bus}}) &= 0 \\ \Delta Q(|V_{\mathrm{bus}}|, \delta_{\mathrm{bus}}) &= 0. \end{aligned} \tag{1.75}$$

Therefore, the mismatch expressions at power frequency for linear PQ buses are given by (1.76). Note that the superscript (1) denotes the parameters at the fundamental frequency.

$$\begin{aligned} \Delta P_i^{(1)} &= P_i^{(1)} - |Y_{ii}^{(1)}||v_i^{(1)}||v_i^{(1)}|\cos(-\theta_{ii}^{(1)}) \\ &\quad - \sum_{j\neq i}|Y_{ij}^{(1)}||v_j^{(1)}||v_i^{(1)}|\cos(-\theta_{ij}^{(1)} - \delta_j^{(1)} + \delta_i^{(1)}) \\ \Delta Q_i^{(1)} &= Q_i^{(1)} - |Y_{ii}^{(1)}||v_i^{(1)}||v_i^{(1)}|\sin(-\theta_{ii}^{(1)}) \\ &\quad - \sum_{j\neq i}|Y_{ij}^{(1)}||v_j^{(1)}||v_i^{(1)}|\sin(-\theta_{ij}^{(1)} - \delta_j^{(1)} + \delta_i^{(1)}). \end{aligned} \tag{1.76}$$

Also, for linear PQ buses, the current at harmonic frequencies is calculated and written as a mismatch expression given by (1.77)

$$I_{\mathrm{bus}}^{(h)} = Y_{\mathrm{bus}}^{(h)} V_{\mathrm{bus}}^{(h)} \tag{1.77}$$

and the harmonic voltages at the buses are given by (1.78)

$$v_i^{(h)} = \overline{z_i^{(h)}}\, i_i^{(h)}, \tag{1.78}$$

where $\overline{z_i}$ is the primitive equivalent impedance at bus i.

If the load at the given bus has a leading power factor, the primitive impedance of the bus is assumed as capacitive, while if the load has lagging power factor the bus impedance is considered reactive. If the load is resistive, the primitive impedance is resistive and frequency independent.

Nonlinear PQ Bus

Next, the power mismatch equations for nonlinear PQ buses are derived. Note that the nonlinear buses result in distorted currents even when the supply voltage is sinusoidal. These are load buses connected to nonlinear loads, such as rectifiers, inverters, and fluorescent lamps. In this section, only rectifier buses are discussed. The expression can be extended to inverter buses, as inverters, and rectifiers have similar voltage and current characteristics. The modeling of buses connected to other nonlinear loads such as fluorescent lamps and gas discharge lights has been done and implemented in commercial software but is not discussed here.

Given the voltage and the firing angle for a power system bus connected to a converter (rectifier), the time domain current can be calculated using the method discussed in Sect. 1.5.2. The calculated time domain current can be resolved into its harmonic components using the Fourier transform. The relationship between the harmonic components of the bus currents and voltage and the firing angle is given by

$$i_i^{(h)} = f_1(v_i^{(h)}, \alpha) \qquad h = 1, 2, 3, \ldots, H, \tag{1.79}$$

where $h = 1$ is the fundamental frequency and $h = H$ is the highest harmonic included in the study.

Also, the specified active power $P_i^{(1)}$ is given by (1.80).

$$P_i^{(1)} = f_2(v_i^{(h)}, \alpha) \qquad h = 1, 2, 3, \ldots, H. \tag{1.80}$$

Similar to conventional power flow, the fundamental frequency power mismatch equations are given by (1.81).

$$\begin{aligned} \Delta P_i^{(1)} &= 0 \\ \Delta Q_i^{(1)} &= 0. \end{aligned} \tag{1.81}$$

The harmonic current continuity expression is written as

$$I_{\text{bus}}^{(h)} = Y_{\text{bus}}^{(h)} V_{\text{bus}}^{(h)} \quad h = 1, 2, 3, \ldots, H. \tag{1.82}$$

Linear PV Bus

At PV or generation buses, fundamental voltage magnitude is known. Therefore, similar to conventional power flow at fundamental frequency, the power mismatch equation is given by (1.83).

$$\Delta P_i^{(1)} = 0. \tag{1.83}$$

Next, power mismatch formulation for the PV buses under harmonic frequency is discussed. To formulate the power flow expression for harmonic frequencies, it is required to understand the behavior of a generator under harmonic voltages. When harmonic voltages are applied to the generator terminals, asynchronous rotating MMF waves are generated. At harmonic frequencies, the resulting MMF waves, which are rotating with respect to the rotor, will produce zero average torque on the rotor. However, these MMF waves will generate currents on rotor iron and rotor windings. As a result, the net equivalent impedance seen by the stator will be the negative sequence impedance of the machine, appropriately scaled for the frequency. Therefore, while modeling PV buses for harmonic frequencies, an equivalent bus to ground impedance is assumed. These impedances are referred to as *harmonic only impedances*, as they only exist for $h > 1$. Based on this discussion, an expression modeling harmonic only impedances of the PV bus is given in (1.84).

$$v_i^{(h)} - \overline{z_i^{(h)}} i_i^{(h)} = 0. \tag{1.84}$$

Finally, the current continuity expression for the PV buses is given by (1.85).

$$I_{\text{bus}}^{(h)} = Y_{\text{bus}}^{(h)} V_{\text{bus}}^{(h)} \quad h = 1, 2, 3, \ldots, H. \tag{1.85}$$

To facilitate the power flow algorithm, the number and type of equations for each bus type are summarized in Table 1.4. The table also presents the number of variables for each equation type. Furthermore, the type and number of state variables are also summarized in Table 1.5. From Tables 1.4 and 1.5, it can be concluded that the problem is well posed, as the number of unknowns and equations are the same. The total number of variables in the study are given by (1.86).

$$\text{Total variables} = 4Hn_{lpq} + (4H + 1)n_{nlpq} + (4H - 3)n_{l\,pv}, \tag{1.86}$$

where H is the total number of frequencies under study. Also, n_{lpq}, n_{nlpq}, and n_{lpv} represent the number of linear PQ, nonlinear PQ, and linear PV buses, respectively.

Table 1.4 Harmonic power flow—types of equations

Bus type	Equation type	Equation form	Valid for harmonic h	Number of equations
LPQ	Power mismatch	$\Delta P^{(1)} = 0$ $\Delta Q^{(1)} = 0$	1	$2n_{\text{lpq}}$
	Current continuity	$I_{bus}^{(h)} = Y_{bus}^{(h)} V_{bus}^{(h)}$	$1 < h \leq H$	$2(H-1)n_{\text{lpq}}$
	Load impedance	$v_i^{(h)} - \overline{z_i^{(h)}}\, i_i^{(h)} = 0$	$1 < h \leq H$	$2(H-1)n_{\text{lpq}}$
NLPQ	FFT of converter current	$i_i^{(h)} - f_1(v_i^{(h)}, \alpha_i) = 0$	$1 < h \leq H$	$2Hn_{\text{nlpq}}$
	Converter power calculation	$P_i^{(1)} - f_2(v_i^{(h)}, \alpha_i) = 0$	1	n_{nlpq}
	Power mismatch	$\Delta P^{(1)} = 0$ $\Delta Q^{(1)} = 0$	1	$2n_{\text{nlpq}}$
	Current continuity	$I_{bus}^{(h)} = Y_{bus}^{(h)} V_{bus}^{(h)}$	$1 < h \leq H$	$2(H-1)n_{\text{nlpq}}$
LPV	Power mismatch	$\Delta P^{(1)} = 0$	1	n_{lpv}
	Harmonic only impedance	$v_i^{(h)} - \overline{z_i^{(h)}}\, i_i^{(h)} = 0$	$1 < h \leq H$	$2(H-1)n_{\text{lpv}}$
	Current continuity	$I_{bus}^{(h)} = Y_{bus}^{(h)} V_{bus}^{(h)}$	$1 < h \leq H$	$2(H-1)n_{\text{lpv}}$

Table 1.5 Harmonic power flow—state variables

State variable	LPQ bus	NLPQ bus	LPV bus
$\lvert V^{(1)} \rvert$	n_{lpq}		
$\delta^{(1)}$	n_{lpq}		n_{lpv}
α		n_{lpq}	
$I^{(h)}$	$2(H-1)n_{\text{lpq}}$	$2(H-1)n_{\text{nlpq}}$	$2(H-1)n_{\text{lpv}}$
$V^{(h)}$	$2(H-1)n_{\text{lpq}}$	$2(H-1)n_{\text{nlpq}}$	$2(H-1)n_{\text{lpv}}$
$I^{(1)}$		$2n_{\text{nlpq}}$	
$V^{(1)}$		$2n_{\text{nlpq}}$	
Total	$4Hn_{\text{lpq}}$	$(4H+1)n_{\text{nlpq}}$	$(4H-3)n_{\text{lpv}}$

Jacobian Matrix

For the fundamental frequency, the Jacobian matrix J_f is of the same format as the conventional power flow study.

$$J_f = \begin{bmatrix} \dfrac{\partial(\Delta P)}{\partial \delta} & \dfrac{\partial(\Delta P)}{\partial |V|} \\ \dfrac{\partial(\Delta Q)}{\partial \delta} & \dfrac{\partial(\Delta Q)}{\partial |V|} \end{bmatrix} = \begin{bmatrix} J_1 & J_2 \\ J_3 & J_4 \end{bmatrix}. \tag{1.87}$$

For the harmonic case, the Jacobian matrix is J_h,

$$J_h = \begin{bmatrix} J_1 & J_2 & J_3 \\ J_4 & J_5 & J_6 \\ J_7 & J_8 & J_9 \end{bmatrix}, \tag{1.88}$$

where the rows of the submatrices correspond to the equation for the three bus types and the columns represent the state variables associated with the equations.

For example, submatrix J_1 is comprised of four row types: active and reactive power mismatch at fundamental frequency and harmonic current and voltage mismatch. Also, submatrix J_1 has four types of rows corresponding to four state variables: fundamental bus voltage magnitude, fundamental bus voltage phase angle, harmonic voltage, and harmonic current. Submatrix J_1 is a square matrix of $4Hn_{\text{lpq}}$ by $4Hn_{\text{lpq}}$. All submatrices corresponding to the harmonic Jacobian matrix J_h with the associated state variables and equation types are summarized in Table 1.6.

Newton–Raphson Method

In previous sections, the harmonic power flow formulation, state variables, mismatch equations, and Jacobian matrix formation were discussed. In this section, the approach to solve the formulated problem using the Newton–Raphson method is summarized. First, the power flow solutions at the fundamental frequency are obtained, using the Newton–Raphson method. Next, the harmonic state variables are initialized and harmonic Jacobian matrix (J_h) is calculated. To avoid matrix inversion, the Jacobian matrix is factored using *LU* decomposition.

$$J_h = (L)(U). \tag{1.89}$$

Next, the update formula given in (1.90) obtained using the factorized harmonic Jacobian matrix is used to update the state variables.

$$\begin{aligned} \Delta F &= -J_h \Delta X \\ &= -(L)(U)\Delta X \\ &= -(L)(W), \end{aligned} \tag{1.90}$$

where ΔF is the vector representing the mismatch in expressions and ΔX is the vector representing the corrections required to update the state variables.

First, vector W is calculated using backward substitution. Next, forward substitution is used to calculate the correction vector ΔX.

Finally, the state variables are updated using (1.91).

$$X_{k+1} = X_k + \Delta X. \tag{1.91}$$

The update process is repeated until the mismatch vector in the power flow equations is sufficiently small. The flowchart for the harmonic power flow solution using the Newton–Raphson method is given in Fig. 1.20.

Table 1.6 Harmonic power flow—Jacobian matrix

	Rows			Columns		
J	Equation type	Bus type	Number of equations	State variable	Bus type	Number of variables
J_1	$\Delta P^{(1)}$	LPQ	n_{lpq}	$\lvert V^{(1)} \rvert$	LPQ	n_{lpq}
	$\Delta Q^{(1)}$	LPQ	n_{lpq}	$\delta^{(1)}$	LPQ	n_{lpq}
	$\Delta I^{(h)}$	LPQ	$2(H-1)n_{\text{lpq}}$	$I^{(h)}$	LPQ	$2(H-1)n_{\text{lpq}}$
	$\Delta V^{(h)}$	LPQ	$2(H-1)n_{\text{lpq}}$	$V^{(h)}$	LPQ	$2(H-1)n_{\text{lpq}}$
J_4	$\Delta I_c^{(h)}$	NLPQ	$2Hn_{\text{nlpq}}$	$\lvert V^{(1)} \rvert$	LPQ	n_{lpq}
	$\Delta P_c^{(h)}$	NLPQ	n_{nlpq}	$\delta^{(1)}$	LPQ	n_{lpq}
	$\Delta P^{(1)}$	NLPQ	n_{nlpq}	$I^{(h)}$	LPQ	$2(H-1)n_{\text{lpq}}$
	$\Delta Q^{(1)}$	NLPQ	n_{nlpq}	$V^{(h)}$	LPQ	$2(H-1)n_{\text{lpq}}$
	$\Delta I^{(h)}$	NLPQ	$2(H-1)n_{\text{nlpq}}$			
J_7	$\Delta P^{(1)}$	LPV	n_{lpv}	$\lvert V^{(1)} \rvert$	LPQ	n_{lpq}
	$\Delta V^{(h)}$	LPV	$2(H-1)n_{\text{lpv}}$	$\delta^{(1)}$	LPQ	n_{lpq}
	$\Delta I^{(h)}$	LPV	$2(H-1)n_{\text{lpv}}$	$I^{(h)}$	LPQ	$2(H-1)n_{\text{lpq}}$
				$V^{(h)}$	LPQ	$2(H-1)n_{\text{lpq}}$
J_2	$\Delta P^{(1)}$	LPQ	n_{lpq}	α	NLPQ	n_{nlpq}
	$\Delta Q^{(1)}$	LPQ	n_{lpq}	$I^{(h)}$	NLPQ	$2(H-1)n_{\text{nlpq}}$
	$\Delta I^{(h)}$	LPQ	$2(H-1)n_{\text{lpq}}$	$V^{(h)}$	NLPQ	$2(H-1)n_{\text{nlpq}}$
	$\Delta V^{(h)}$	LPQ	$2(H-1)n_{\text{lpq}}$	$I^{(1)}$	NLPQ	$2n_{\text{nlpq}}$
				$V^{(1)}$	NLPQ	$2n_{\text{nlpq}}$
J_5	$\Delta I_c^{(h)}$	NLPQ	$2Hn_{\text{nlpq}}$	α	NLPQ	n_{nlpq}
	$\Delta P_c^{(h)}$	NLPQ	n_{nlpq}	$I^{(h)}$	NLPQ	$2(H-1)n_{\text{nlpq}}$
	$\Delta P^{(1)}$	NLPQ	n_{nlpq}	$V^{(h)}$	NLPQ	$2(H-1)n_{\text{nlpq}}$
	$\Delta Q^{(1)}$	NLPQ	n_{nlpq}	$I^{(1)}$	NLPQ	$2n_{\text{nlpq}}$
	$\Delta I^{(h)}$	NLPQ	$2(H-1)n_{\text{nlpq}}$	$V^{(1)}$	NLPQ	$2n_{\text{nlpq}}$
J_8	$\Delta P^{(1)}$	LPV	n_{lpv}	α	NLPQ	n_{nlpq}
	$\Delta V^{(h)}$	LPV	$2(H-1)n_{\text{lpv}}$	$I^{(h)}$	NLPQ	$2(H-1)n_{\text{nlpq}}$
	$\Delta I^{(h)}$	LPV	$2(H-1)n_{\text{lpv}}$	$V^{(h)}$	NLPQ	$2(H-1)n_{\text{nlpq}}$
				$I^{(1)}$	NLPQ	$2n_{\text{nlpq}}$
				$V^{(1)}$	NLPQ	$2n_{\text{nlpq}}$
J_3	$\Delta P^{(1)}$	LPQ	n_{lpq}	$\delta^{(1)}$	LPV	n_{lpv}
	$\Delta Q^{(1)}$	LPQ	n_{lpq}	$I^{(h)}$	LPV	$2(H-1)n_{\text{lpv}}$
	$\Delta I^{(h)}$	LPQ	$2(H-1)n_{\text{lpq}}$	$V^{(h)}$	LPV	$2(H-1)n_{\text{lpv}}$
	$\Delta V^{(h)}$	LPQ	$2(H-1)n_{\text{lpq}}$			
J_6	$\Delta I_c^{(h)}$	NLPQ	$2Hn_{\text{nlpq}}$	$\delta^{(1)}$	LPV	n_{lpv}
	$\Delta P_c^{(h)}$	NLPQ	n_{nlpq}	$I^{(h)}$	LPV	$2(H-1)n_{\text{lpv}}$
	$\Delta P^{(1)}$	NLPQ	n_{nlpq}	$V^{(h)}$	LPV	$2(H-1)n_{\text{lpv}}$
	$\Delta Q^{(1)}$	NLPQ	n_{nlpq}			
	$\Delta I^{(h)}$	NLPQ	$2(H-1)n_{\text{nlpq}}$			

(continued)

Table 1.6 (continued)

	Rows			Columns		
J	Equation type	Bus type	Number of equations	State variable	Bus type	Number of variables
J_9	$\Delta P^{(1)}$	LPV	n_{lpv}	$\delta^{(1)}$	LPV	n_{lpv}
	$\Delta V^{(h)}$	LPV	$2(H-1)n_{\mathrm{lpv}}$	$I^{(h)}$	LPV	$2(H-1)n_{\mathrm{lpv}}$
	$\Delta I^{(h)}$	LPV	$2(H-1)n_{\mathrm{lpv}}$	$V^{(h)}$	LPV	$2(H-1)n_{\mathrm{lpv}}$

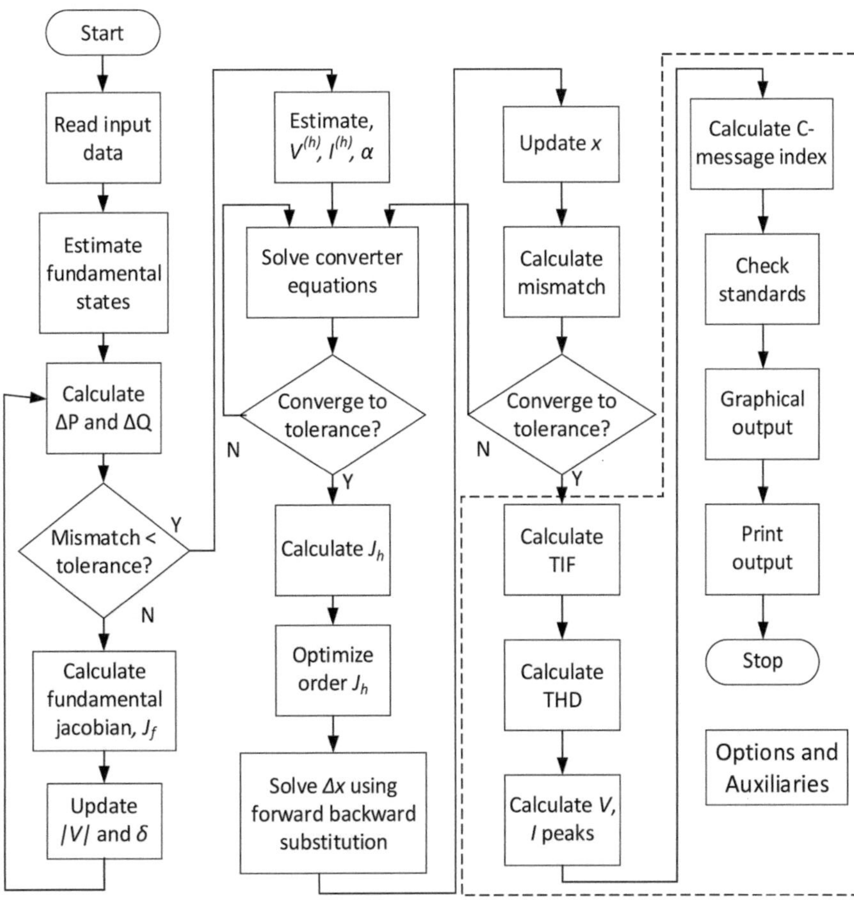

Fig. 1.20 Newton–Raphson harmonic power flow study

The concluding remarks for Newton–Raphson based harmonic power flow are summarized as follows. Note that a Newton–Raphson based harmonic power flow study can take about 20 iterations compared to 2–3 iterations in the Newton–Raphson based conventional power flow.

- The Jacobian matrix for the harmonic power flow is much larger than the conventional power flow case and usually contains H times more rows and columns, where H is the number of harmonic frequencies included in the study.
- Since the harmonic power flow requires to solve for the converter current within each iteration of the power flow algorithm, the execution time for harmonic power flow is significantly longer.
- However, since the Jacobian matrix (for harmonic case as well) is sparse, the memory requirement is not severe.
- Unlike conventional power flow, the harmonic power flow does not model several characteristics, such as tap changing load transformers, reactive power limits, and other features of conventional power flow. Therefore, harmonic power flow is not a replacement for the conventional power flow. Instead, these characteristics are not included in the harmonic power flow study in order to simplify the algorithm and focus on the harmonic signal levels.

1.6 Harmonic Filters

As discussed in the previous sections, due to nonlinear loads and power electronic switching, the voltage and current at the point of connection to the circuit are not purely sinusoidal and can contain a considerable amount of high frequency components. If these non-sinusoidal voltages or currents are injected into the distribution system, they will distort the supply voltages of other devices and will increase the system power losses. Filters, therefore, are typically implemented in the circuit to prevent harmonic currents from flowing into the distribution system, aiming to keep the system THD below 5 % [10]. Figure 1.21 presents input current

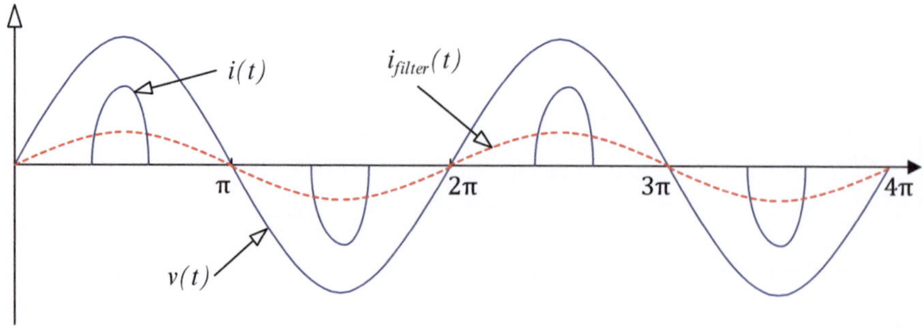

Fig. 1.21 Current waveform without filter and with filter

waveforms of a rectifier without and with a filter. The rectifier current waveform without a harmonic filter is not a sine wave and contains a significant amount of harmonics amounting to current THD up to 110 %. A harmonic filter reconstructs the non-sinusoidal current waveform into a sinusoidal waveform and thus significantly reduces the system THD.

Harmonic filters are usually designed to shunt the flow of harmonic currents. Their working principle is typically based on the tuned resonance between inductor and capacitor. The filter's inductor and capacitor elements result in a resonance effect and provide a low impedance path for harmonic currents, resulting in a near sinusoidal waveform at the fundamental frequency.

Harmonic filters are primarily categorized as *active* or *passive* filters. An active filter is a combination of amplifiers or transistors, resistors, and capacitors. The gain of this filter can be any arbitrary value depending on the designed line impedance. A passive filter, on the other hand, uses only passive components such as inductors and capacitors to remove the harmonics. This type of filter produces very little noise, is rugged to the environment, and is easy to implement. Therefore, passive filters are preferred over active filters to mitigate power system harmonic concerns.

1.6.1 Active Harmonic Filter

An active filter can detect harmonic components and produce a counteracting signal to partially or totally eliminate harmonics. The filter is constructed in a way that high frequency signals such as fifth and seventh order harmonics can be identified. For example, if a bus contains 20 % of fifth order harmonic voltage components, a controller can be implemented in the filter to check for the fifth order harmonics and generate an identical but 180° out of phase signal to cancel it.

Figure 1.22 shows four typical types of active filters: series voltage, series current, parallel voltage, and parallel current [1]. In this section, the series voltage type (see Fig. 1.22a) is examined to understand the operating principle of a general active filter. The distorted voltage $V_{\text{in}}(t)$ with harmonic contents is supplied as the filter input. The detector, which consists of a potential transformer and a high pass filter, blocks the fundamental frequency and passes the harmonic signals $V_{af}(t)$. The harmonic voltages are then added back to the input voltage via a series inversed polarity transformer. The filter output voltage is given by

$$V_{out}(t) = V_{in}(t) + kV_{af}(t). \tag{1.92}$$

The value of k depends on many factors, such as transformer ratio and amplifier gain, and thus needs to be designed carefully. The drawbacks of this filter are the limited bandwidth with undesired attenuation, phase shift, and delay.

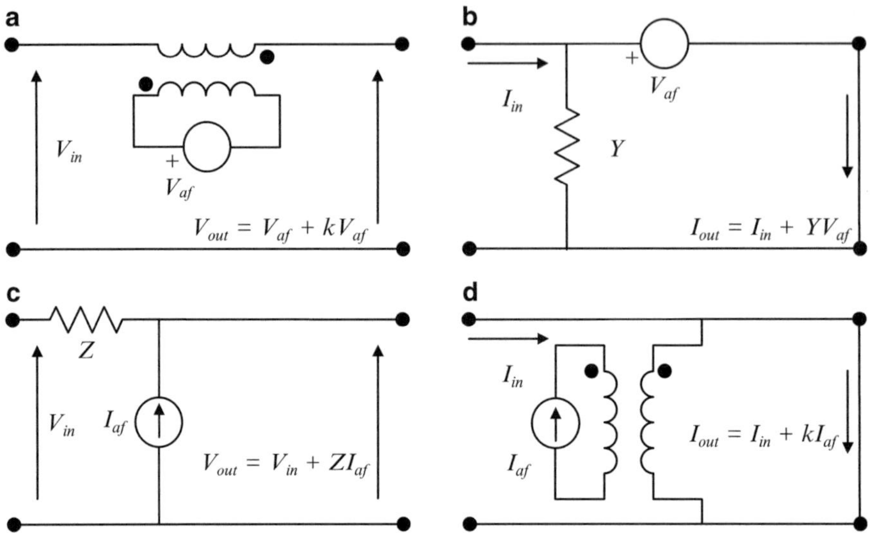

Fig. 1.22 Typical active filter types: (**a**) Series voltage, (**b**) Series current, (**c**) Parallel voltage, and (**d**) Parallel current

Active filters have been successfully implemented in low power levels such as audio and communication applications. However, at high power levels the situation is different. The filter requires a complicated design process and special power circuit components, thus making it less popular for power system applications.

1.6.2 *Passive Harmonic Filter*

Unlike active filters, passive filters cannot generate gain higher than 1. However, these filters require no power supplies to operate and thus can work well at high frequency. Passive filters are further classified into four types: low-pass filter, high-pass filter, band-stop filter, and band-pass filter [11]. These filters are defined as follows:

1. Low-pass filter—This filter, as shown in Fig. 1.23a, is characterized by its low impedance value under a specific cut-off frequency f_c (−3 dB) and high impedance value above that frequency. A low-pass filter, therefore, passes through all signals under f_c and attenuates all signals above f_c.
2. High-pass filter—A high-pass filter has the opposite effect of a low-pass filter. It blocks all signals below the cut-off frequency f_c due to high impedance value and passes all high frequency signals (see Fig. 1.23b).

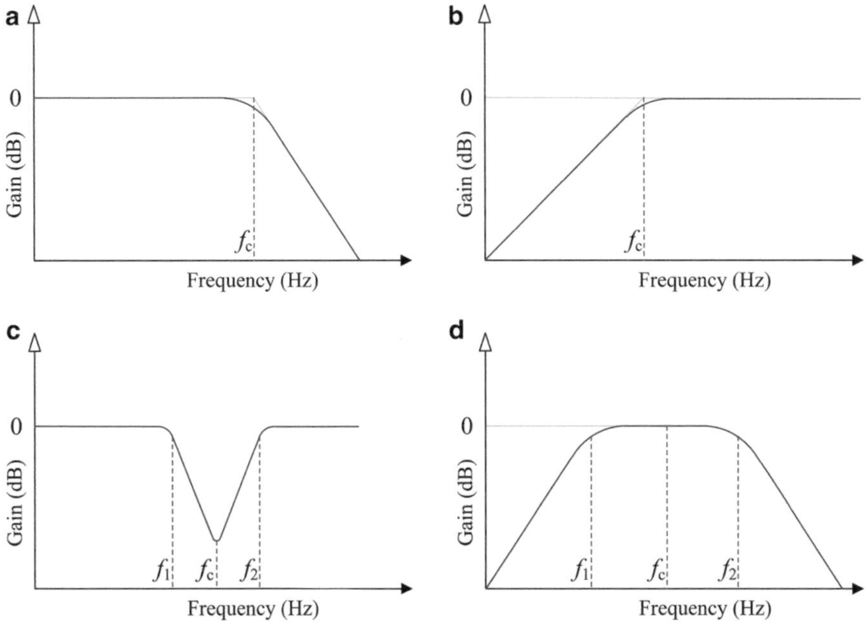

Fig. 1.23 Passive harmonic filter type: (**a**) Low-pass filter, (**b**) high-pass filter, (**c**) band-stop filter, and (**d**) band-pass filter

3. Band-stop filter—A band-stop or notch filter passes all signal frequencies except for a very narrow band (see Fig. 1.23c). The filter gain around f_c is much lower than the rest of the frequency band, and thus the amplitude of the signal within the narrow frequency band is significantly reduced.
4. Band-pass filter—This filter has the reverse attenuation behavior compared to the band-stop filter. The filter gain is near 1 for a specific range of frequencies $[f_1 - f_2]$ and zero for the rest (see Fig. 1.23d). The harmonics outside the band-pass range are therefore eliminated from the signal.

The harmonic content resulting from the switching of semiconductor devices in converters, for instance, has frequencies in the range of several kHz. Because filters are designed to attenuate signals above a specific frequency, low-pass filters are normally deployed. Several configurations of low-pass filters are presented in the following section.

Low-Pass Filter Configurations

As mentioned earlier, the low-pass filter utilizes passive components such as inductors and capacitors to eliminate high frequency harmonic content above the cut-off frequency. The filter attenuation capability depends on the slope or stiffness of filter after the cut-off frequency f_c. Figure 1.24 presents three common configurations of filter in power circuits: L-filter, LC-filter, and LCL-filter.

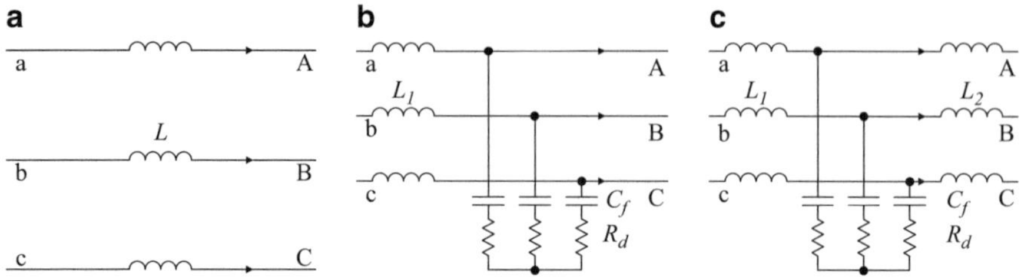

Fig. 1.24 Common low-pass filter configurations: (**a**) L-filter, (**b**) LC-filter, and (**c**) LCL-filter

L-Filter

An L-filter or choke is the simplest configuration with only an inductor connected between the input and the output terminals. Figure 1.24a shows an L-filter connected to a three-phase system. The transfer function for the filter is derived from the output and input relationship as follows:

$$G(s) = \frac{\text{Output}(s)}{\text{Input}(s)} = \frac{1}{Ls}. \tag{1.93}$$

From (1.93), the L-filter is characterized as a first-order system with the ability to attenuate harmonics at 20 dB/dec over the complete frequency range, above the cut-off frequency. Therefore, this filter is applicable for systems with high frequencies resulting from components such as switching power converters. However, in order to have sufficient attenuation, the size of an L-filter should be adequately large. This results in an increase in the total size and cost of the system. More importantly, adding a large inductance into the power system can affect and degrade system's dynamic characteristics.

LC-Filter

An LC-filter is developed from an L-filter by inserting a capacitor in parallel right after the inductor terminal as shown in Fig. 1.24b. The LC-filter has better performance compared to the L-filter because it is a second-order system with transfer function given by

$$G(s) = \frac{\text{Output}(s)}{\text{Input}(s)} = \frac{\omega_{\text{res}}^2}{(s^2 + (R_d/L_1)s + \omega_{\text{res}}^2)}, \tag{1.94}$$

where $\omega_{\text{res}}^2 = 1/(L_1 C_f)$ and R_d is the passive damping resistor connected in series with C_f in order to damp the resonance in the circuit.

As with the L-filter, an LC-filter passes all low frequency signals below the cut-off frequency. However, this filter has a better response to high frequency harmonic current; its attenuation ability to eliminate harmonics above the resonant frequency, ω_{res}, is 40 dB/dec. In the LC-filter, the harmonic components are trapped into the inductor and capacitor. The inductance, therefore, can be significantly reduced by adding more capacitance. The filter cost and loss, thus, can be scaled down. Practically, however, the capacitor should store less than 5 % of power passing through it. Otherwise, it can result in a high value of capacitance, leading to a high inrush current at the fundamental frequency.

LCL-Filter

Figure 1.24c presents the configuration of a three-phase LCL-filter. This filter has one more inductor compared to the LC-filter: an inductor L_1 at the input side, an inductor L_2 at the output, and a capacitor placed in between. The currents first pass through L_1 where most harmonic components are trapped. Inductor L_2 and the capacitor then diminish the current ripples at switching frequencies to a desired level. The LCL-filter transfer function can be given from direct relationship between output and input in Fig. 1.25c:

$$G(s) = \frac{i(s)}{v(s)} = \frac{R_d}{L_2 L_1 s} \frac{(s + Z_{RC})}{(s^2 + R_d C_f \omega_{\text{res}}^2 s + \omega_{\text{res}}^2)}, \tag{1.95}$$

where $Z_{LC}^2 = 1/(L_2 C_f), Z_{RC}^2 = 1/(R_d C_f), \text{and} \omega_{\text{res}}^2 = (L_1 + L_2/L_1) Z_{LC}^2$.

The LCL-filter from (1.95) is modeled as a third-order system with the attenuation ability of 60 dB/dec. This means that the filter can reduce the amplitudes of harmonic components above the cut-off frequency to nearly zero. In addition, the LCL-filter supports the system robustness in the case of variations in the output impedance. For example, considering a change at the output side impedance:

$$\omega_{\text{res1}}^2 = \frac{L_1 + L_2 + L_{g1}}{L_1(L_2 + L_{g1})C_f} \text{ and } \omega_{\text{res2}}^2 = \frac{L_1 + L_2 + L_{g2}}{L_1(L_2 + L_{g2})C_f}, \tag{1.96}$$

where grid impedance L_{g1} changes to L_{g2}.

The frequency variation can be estimated as

$$\begin{aligned} \omega_{\text{res1}}^2 - \omega_{\text{res2}}^2 &= (\omega_{\text{res1}} - \omega_{\text{res2}})(\omega_{\text{res1}} + \omega_{\text{res2}}) \approx 2\omega_{\text{res}} \Delta\omega_{\text{res}} \\ \Delta\omega_{\text{res}} &= \frac{1}{2\omega_{\text{res}} C_f} \left(\frac{1}{L_2 + L_{g1}} - \frac{1}{L_2 + L_{g2}} \right). \end{aligned} \tag{1.97}$$

It is clear from (1.97) that the impedance variation can be significantly reduced by using a large capacitor. The trade-off for an LCL-filter compared to the two previous L- and LC-filters is the complexity of the configuration. The transfer function (1.95) indicates that an LCL-filter adds one more zero and two more poles than an L-filter which can cause unstable states in the system due to resonances. The LCL-filter thus requires a careful design process before installing in the power system.

In summary, each of the three filters has advantages, and depending on the power system application, one can decide which filter should be installed. In electric machine drives, for instance, an L-filter can remove most harmonic components. In grid-connected applications, however, the quality requirements of current injected into the grid are high. An LC- or LCL-filter, therefore, is deployed to comply with grid codes. Moreover, harmonic filters need to be carefully designed to ensure both harmonic cancelation capability and system stability. In practice, a low-pass filter is usually combined with a current or voltage controller in order to have a clean and smooth output signal while ensuring system stability.

1.7 Conclusion

This chapter presents a thorough discussion on distribution circuit modeling, operation, and analysis in the presence of power system harmonics. The method to mitigate harmonic concerns using harmonic filters is also reviewed. The chapter is written in honor of Dr. Heydt for his seminal work in power quality, particularly in power system harmonics and harmonic power flow study. Therefore, most of the material presented in this chapter is a summary of his research and contributions pertaining to the subject.

First, the topic of power system harmonics is introduced and its impacts on the distribution circuit are explained. Next, several measures to quantify power system harmonics are defined, followed by a comprehensive discussion on major sources of harmonics and an approach to model nonlinear loads and network components under non-sinusoidal operating conditions. The method for distribution circuit analysis in the presence of power system harmonics is also reported. For this purpose, several algorithms for load flow analysis under non-sinusoidal operating conditions, i.e., harmonic power flow algorithms are reviewed. Finally, the application of harmonic filters in mitigating power system harmonic concerns is presented, including a detailed account on modeling, analysis, and utility of both active and passive filters.

Acknowledgments The authors would like to acknowledge the contribution of Tuan Ngo and Min Lwin in preparing this chapter.

References

1. Heydt GT (1994) Electric power quality, 2nd edn. Stars in a Circle Publication, West Lafayette
2. Xia D, Heydt G (1982) Harmonic power flow studies Part I - formulation and solution. IEEE Trans Power Apparatus Syst PAS 101(6):1257–1265
3. Xia D, Heydt G (1982) Harmonic power flow studies Part II - implementation and practical application. IEEE Trans Power Apparatus Syst PAS 101(6):1266–1270
4. Song W, Heydt G, Grady W (1984) The integration of HVDC subsystems into the harmonic power flow algorithm. IEEE Trans Power Apparatus Syst PAS 103(8):1953–1961

5. Gotham D, Heydt G (1998) Power flow control and power flow studies for systems with FACTS devices. IEEE Trans Power Syst 13(1):60–65
6. Heydt G (1989) Identification of harmonic sources by a state estimation technique. IEEE Trans Power Deliv 4(1):569–576
7. Beides H, Heydt G (1991) Dynamic state estimation of power system harmonics using Kalman filter methodology. IEEE Trans Power Deliv 6(4):1663–1670
8. Najjar M, Heydt G (1991) A hybrid nonlinear-least squares estimation of harmonic signal levels in power systems. IEEE Trans Power Deliv 6(1):282–288
9. Santoso S (2012) Fundamentals of electric power quality. CreateSpace, Austin
10. IEEE Recommended Practices and Requirements for Harmonic Control in Electrical Power Systems (1993) IEEE Std 519-1992, April 1993, pp 1–112
11. Teodorescu R, Liserre M, Rodriguez P (2011) Grid converter for photovoltaic and wind power systems. Willey, Chichester

Chapter 2
A Meta-heuristic Approach for Optimal Classification of Power Quality Disturbances

B.K. Panigrahi and Nilanjan Senroy

2.1 Introduction

In recent years, power quality has become a significant issue worldwide. The proliferation of solid state switching devices, nonlinear loads, unbalanced network conditions, power electronics based lighting control systems, switched mode power supplies, and industrial rectifiers and inverters have resulted in complicated power quality issues [1]. These include quasi-static harmonic dynamic voltage distortions, inrush, pulse type current phenomena with excessive harmonics and highly distorted line voltage and current waveforms. Other power quality events involve variations in the electricity supply, such as voltage dips and fluctuations, momentary interruptions, harmonics and oscillatory transients causing failure or mal-operation of the equipment. Improvements in power quality must start with fast and reliable detection of the disturbances and the sources and causes of such disturbances must be known before any appropriate mitigating action can be taken.

In order to determine the causes and sources of disturbances, one must have the ability to detect and localize these disturbances. The wavelet transform (WT) [2–4] is widely used in analyzing non-stationary signals for power quality assessment [5, 6]. In order to identify the type of disturbance present in a measured voltage/current signal, different methodologies based on combination of the wavelet transform and artificial neural networks (ANNs) have been explored [7]. Using the multi-resolution properties of WT [6], the features of the disturbance signal are extracted at different resolution levels. These features are used to train different

B.K. Panigrahi • N. Senroy (✉)
Department of Electrical Engineering, Indian Institute of Technology Delhi,
New Delhi 110016, India
e-mail: nsenroy@ee.iitd.ac.in

E. Kyriakides et al. (eds.), *Electric Power Engineering Research and Education*,
Power Electronics and Power Systems, DOI 10.1007/978-3-319-17190-6_2

ANN systems, to extract important information from a disturbance signal and determine the type of disturbance that has caused a power quality problem to occur. Wavelets have also been used with probabilistic neural network to classify seven types of power quality (PQ) events [8, 9].

In this chapter the Wavelet Transform is used to extract relevant features from power system disturbance signals. These signals are decomposed to up to 13 levels; seven statistical measures—energy, entropy, standard deviation, mean, kurtosis, Instantaneous Transient Disturbance (ITD) and skewness are defined for each level. For any signal, the total feature set has 91 elements for a 13 level decomposition. Five state-of-the art classifiers namely, linear discriminant analysis (LDA), fuzzy k-nearest neighbor (FKNN), general regression neural network (GRNN), radial basis function (RBF), and probabilistic neural network (PNN) are considered [10]. For classifier combiners, the choice is made between Majority Voting, Borda count, and Bayesian Belief methods [10]. The crucial task of selecting the most appropriate classifiers and combiner, as well as finding the optimal feature set, has been performed using an optimization algorithm which maximizes the accuracy of classification.

The optimization task as outlined above is a nonlinear complex problem with a considerably large dimensionality, necessitating the use of meta-heuristic algorithms. Differential Harmony Search (DHS) [11] is such an algorithm, created by incorporating a difference vector mutation of classical Differential Evolution Algorithm (DE) [12, 13] in the mutation strategy of that of Harmony Search (HS) [14]. The DHS search strategy possesses a considerably high convergence speed. It is proposed to use this algorithm to create an optimal pattern recognition machine for classification of PQ disturbances. The classification accuracy obtained has been compared with various cases to highlight the effectiveness of the procedure.

2.1.1 *Wavelet Transform*

Many signal processing techniques have been successfully applied to extract important time frequency information from the power quality signals. The Wavelet Transform [2, 3] is one of the important signal processing tools for the time-frequency representation of the signal. A detailed description of the WT application for power quality assessment, detection, and classification is well presented in refs. [5–8].

The Discrete Wavelet Transform (DWT) decomposes the non-stationary power quality signal into various frequency sub-bands using low-pass and high-pass filters subsequently. In this work we have considered the db4 wavelet as the mother wavelet for the time-frequency resolution of the signal.

2.1.2 Data Preparation and Feature extraction

Various power quality disturbance signals are simulated using the analytical expression as detailed in refs. [15, 16]. Around 200 signals are generated for each power quality event by varying the parameters describing the particular event. Various sag and swell signals are generated by varying the magnitude and duration of the event. Similarly, harmonic signals are generated with various magnitudes of harmonic content. Sag and swell with harmonics events are created by varying the harmonic number, harmonic content and sag and swell magnitude and duration. Transient signals are generated with various transient frequencies and rate of decay of transients. The sampling frequency used in this work is 6.4 kHz and the data window considered is ten fundamental cycles.

Once the signals are generated, they are processed with the WT to extract various information at various frequency sub-bands. The detailed coefficient D_{ij} at each decomposition level is used to extract the features. Statistical features like energy, standard deviation, mean, kurtosis, skewness, entropy, and Instantaneous Transient Disturbance (ITD) of the decomposition coefficients D_{ij} are calculated by using the following equations

$$\text{Energy ED}_i = \sum_{j=1}^{N} \left|D_{ij}\right|^2 \quad i = 1, 2 \ldots l \tag{2.1}$$

$$\text{Standard Deviation } \sigma_i = \left(\frac{1}{N-1}\sum_{j=1}^{N}\left(D_{ij} - \mu_i\right)^2\right)^{1/2}, \quad i = 1, 2 \ldots l \tag{2.2}$$

$$\text{Mean } \mu_i = \frac{1}{N}\sum_{j=1}^{N} D_{ij} \quad i = 1, 2 \ldots l \tag{2.3}$$

$$\text{Kurtosis KRT}_i = \frac{E\left(D_{ij} - \mu_i\right)^4}{\sigma_i^4} \quad i = 1, 2 \ldots l \tag{2.4}$$

$$\text{Skewness SK}_i = \frac{E\left(D_{ij} - \mu_i\right)^3}{\sigma_i^3} \quad i = 1, 2 \ldots l \tag{2.5}$$

$$\text{Entropy ENT}_i = -\sum_{j=1}^{N} D_{ij}^2 \log\left(D_{ij}^2\right) \quad i = 1, 2 \ldots l \tag{2.6}$$

$$\text{ITD}_i = \frac{\text{ED}_i}{\text{Energy of Fundamental}} \tag{2.7}$$

where $i = 1,2,\ldots,l$ is the wavelet decomposition level from level 1 to level l. $j = 1,2\ldots N$, where N is the number of coefficients of detail at each decomposition level "l". The ITD is a measure of the ratio of the energy of the disturbance which is calculated from all the continuous frequencies (not limited to harmonics only) to the fundamental energy.

Thus in the present case, with a "13" level decomposition the feature vector adopted is of length "7×13" i.e., 91. The feature vector may thus be denoted as

$$\begin{aligned} \text{Feature} = \big[&\text{ED}_1 \ \text{ED}_2 \ldots \text{ED}_l \ \sigma_1 \ \sigma_2 \ldots \sigma_l \mu_1 \ \mu_2 \ldots \mu_l \\ &\text{KRT}_1 \ \text{KRT}_2 \ldots \text{KRT}_l \ \text{SK}_1 \ \text{SK}_2 \ldots \text{SK}_l \\ &\text{ENT}_1 \ \text{ENT}_2 \ldots \text{ENT}_l \ \text{ITD}_1 \ \text{ITD}_2 \ldots \text{ITD}_l\big] \end{aligned} \tag{2.8}$$

2.2 Classifiers and Combiners

2.2.1 Classifiers

Using the feature extraction process described in the previous section, each data is mapped to a 91-dimensional space. This processed set is available to every classifier and an optimal set is used for classification. The classifiers used include both simple classifiers like LDA and FKNN, as well as complex classifiers based on neural networks like RBF, GRNN, and PNN. A detailed description of these classifiers is available in ref. [10] for the interested reader. The output from these classifiers is modified so that hard decisions as well as ranking measures are obtained. For classifiers which generally provide ranks of different classes related to a particular data, the highest ranked class is taken as the hard decision for that classifier. On the other hand, for those classifiers providing only hard decisions about the class number of the input data, the rest of the classes are assigned a lowest rank, i.e., if n classes are present and the classifier assigns class i to the data point, then the remaining $n-1$ classes are assigned rank n while the ith class is assigned the rank 1 for the concerned data point. This step ensures compatibility of all the classifiers with combiners working on ranks as well as class decisions from classifiers.

2.2.2 Combiners

Class decisions (hard decision or ranks) are passed to all the combiners. A study of recent research in this area has shown that Majority Voting, Borda Count, and Bayesian Belief algorithms are efficient classifier combiners. While Majority Voting and Bayesian Belief work with hard classification decisions, the Borda Count differentiates ranking measures for every classifier. A detailed description of all these algorithms is available in ref. [10].

2.2.3 *Overview of Optimization algorithm*

2.2.3.1 Harmony Search

Harmony Search (HS) [14] is a meta-heuristic algorithm that mimics the process of improvisation employed by musicians to obtain optimum harmony of musical notes. During the optimization of an engineering problem, the solution is estimated by trying out different values of the decision variables, and calculating the value of the objective/fitness function with respect to various aspects such as cost, efficiency, and accuracy. Just as in musical improvization, the best state is determined by an aesthetic standard, the optimization process also seeks a best state or a global optimum on the basis of the objective function evaluation. Further, the pitch of each musical instrument determines the aesthetic quality of the music produced; similarly the objective function value is determined by the set of values assigned to each decision variable. Finally, just as the aesthetic sound quality is improved by repeated practice, the objective function value can also be improved in an iterative manner. The HS algorithm comprises of the following five steps:

Step 1: Initialization of the Optimization Problem and Algorithm Parameters

In the first step, the optimization problem is specified as follows:

Minimize (or Maximize) $f\left(\vec{x}\right)$

subject to the condition

$$x_i \in X_i, \forall i = 1 \ldots N \tag{2.9}$$

where *f*(.) is a scalar objective function; $\vec{x}$ is a vector comprising of the decision variables x_i; X_i is the set of possible range of values for each decision variable x_i, and N is the number of decision variables. Additionally, the control parameters of HS are the Harmony Memory Size (HMS), i.e., the number of solution vectors (termed as population members) in the harmony memory (for each generation); Harmony Memory Considering Rate (HMCR); Pitch Adjusting Rate (PAR); and the Number of Improvizations (*NI*) which is the stopping criterion.

Step 2: Harmony Memory initialization

In the second step each component of the solution vector in the parent population (Harmony Memory), which is of size HMS, is initialized with a uniformly distributed random number between the upper and lower bounds $[{}_Lx_i,\ {}_Ux_i]$, where $1 \leq i \leq N$. For the *i*th component of the *j*th solution vector, this step can be represented by the following equation:

$$x_i^j = {}_Lx_i + \text{rand}\,(0, 1) \cdot ({}_Ux_i - {}_Lx_i) \tag{2.10}$$

where $j = 1, 2, 3, \ldots,$ HMS and rand (0,1) is a uniformly distributed random number between 0 and 1.

Step 3: New Harmony Improvisation

In this step, a new Harmony vector $\vec{x}' = \left(x_1', x_2', \ldots.. x_N'\right)$ is generated based on three rules: (1) memory consideration, (2) pitch adjustment, and (3) random selection. This process of generating a new Harmony is called "improvization." The steps may be elaborated by the following set of equations:

Memory consideration:

$$x_1' \leftarrow \begin{cases} x_i \in \left\{x_i^1, x_i^2, \ldots.. x_i^{\mathrm{HMS}}\right\} \text{ with probability HMCR} \\ x_i \in X_i \text{ with probability } 1 - \mathrm{HMCR} \end{cases} \tag{2.11}$$

Pitch Adjustment:

$$x_1' = \begin{cases} x_1' \pm \mathrm{rand}(0,1) \cdot \mathrm{bw} \text{ with probability PAR} \\ x_1' \quad \text{with probability } 1 - \mathrm{PAR} \end{cases} \tag{2.12}$$

bw refers to an arbitrary distance bandwidth (a scalar number). Step 3 essentially generates a new variation in the algorithm and is similar to the concept of mutation in the standard evolutionary algorithms. Thus, either the decision variable is perturbed with a random number between 0 and bw, or it is left unaltered with a probability PAR or else it is left unchanged with probability (1 – PAR).

Step 4: Harmony memory update:

If the new Harmony vector $\vec{x}' = \left(x_1', x_2', \ldots.. x_N'\right)$ is better in terms of the objective function value than the worst Harmony in the HM, then the new harmony is included in the HM and the existing worst harmony is excluded from the HM. Conversely, if the new Harmony vector yields an objective function value worse than the worst Harmony, then it is discarded. Thus, the selection process involves the evaluation of the objective function for every new Harmony vector, to determine whether the new variation should be included in the population (Harmony Memory) or not.

Step 5: Check Stopping Criterion

Steps 3 and 4 are repeated until the stopping criterion (maximum number of improvizations) is satisfied.

2.2.4 *Harmony Search with Modified Differential Mutation Operator and its Discrete Version*

Experiments with classical HS meta-heuristics over standard numerical benchmark functions have indicated that the algorithm suffers from the problem of premature and/or false convergence, with slower convergence particularly over a multimodal fitness landscape. Further, the performance of the classical HS algorithm deteriorates with an increase in the dimensionality of the solution space. To overcome these difficulties related with HS, a new variant termed Differential Harmony Search (DHS) was proposed in ref. [11]. In this variation, the explorative capability of HS was improved by replacing the pitch adjustment strategy of the classical version of the algorithm with the mutation scheme employed in DE/rand/1 [12, 13], with corroborative experimental results. When a large region of the solution space is prohibited by security constraints of the problem at hand, the mutation scheme used in DHS is modified by incorporating the one employed in DE/target-to-best/1 [12, 13] and is then modulated by the scaling parameters. Thus, the Harmony Search with Modified Differential mutation operator (HS_MD) follows steps 1–5 as previously outlined for classical HS, with the pitch adjustment operation (Eq. 2.12) in Step 3 being replaced by the following equation:

$$x_1' = x_1' + a_1 \text{*rand*} \left(x_{\text{BEST}} - x_1' \right) + a_2 \text{*rand*} (x_{r1} - x_{r2}) \tag{2.13}$$

where $r_1 \neq r_2$ are two random indices of vectors chosen from the harmonic memory; and a_1, a_2 are two scaling parameters introduced to bias the vectors towards the global optimum. This step ensures that for complex problems with constraint equations, the vectors quickly pass into the "legal" zone and are provided with the opportunity to explore the "legal zone" more effectively. Typical values of a_1, a_2 are 6 and 0.3, respectively.

For the discrete version of DHS, *DHS_MD*, the initialization step is represented by:

$$x_i^j = {}_L x_i + \text{ceil}(\text{rand}(0, 1) \cdot ({}_U x_i - {}_L x_i)) \tag{2.14}$$

and the mutation operation is modified as:

$$x_1' = x_1' + \text{ceil}\ \left(a_1 * \text{rand} * \left(x_{\text{BEST}} - x_1'\right) + \ \text{ceil}\left(a_2 * \text{rand} * (x_{r1} - x_{r2}\right) \quad (2.15)$$

The rest of the steps for the algorithm proceed just as in the classical HS. A flowchart for the algorithm is given in Fig. 2.1.

Fig. 2.1 Flowchart for DHS_MD

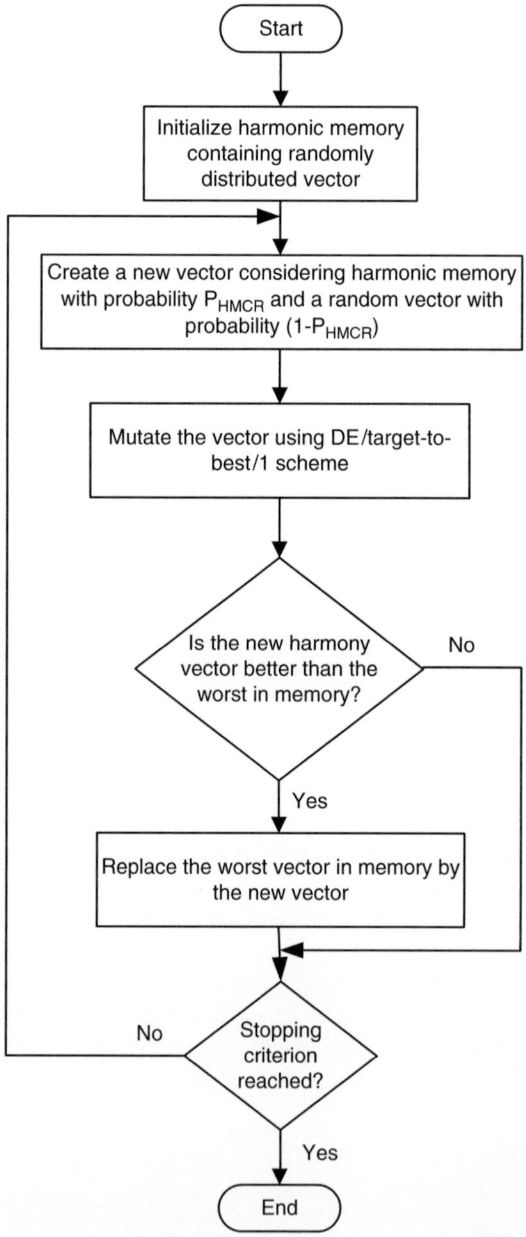

2.2.5 Optimum Classification Using Differential Harmony Search

The previous sections have described the basic framework of how measured PQ data may be processed to extract features. The process of combining classifiers hierarchically to generate classification decisions has also been described. Every measurement yields a data point, and these decisions are obtained for every such data point. If the classes of the data points are known, then by comparing the classifier results with the known classes, an average classification accuracy may be calculated. It is desired that this accuracy is as high as possible. The feature extraction process produces a very high dimensional space—for the present case using DWT up to 13 levels of decomposition and seven statistical measures per level, a 91 dimensional feature space emerges. Such large dimensionality not only hinders the overall computational efficiency, but may also produce aliasing effects due to overlapping features, thus degrading the performance of classifiers. Hence, DHS is expected to provide with a feature space small enough to minimize the computational burden, yet extensive enough so that the various classes are well differentiated within the entire feature. Finally, classification by different techniques must be followed by selecting the correct combiner by the optimization algorithm.

Thus, in order to accommodate all the requirements in a simple yet effective way, each member of the DHS population is selected to be "n" dimensional ($n = m + p + 1$) of which the first "m" dimensions are reserved for the selected features (note: the maximum number of features allowed to be selected is "m"). The next "p" dimensions indicate the selected classifiers, while the last dimension is indicates the combiner selected.

For the feature selection components, the components are allowed to lie in the integral range between 1 and "$N+1$", where "N" is the dimensionality of the feature space produced by the DWT with the value directly representing the feature selected. If any component magnitude is "$N+1$", the interpretation is that no feature is selected. This process ensures that DHS can be supplied with a feature space lower than m-dimension if that can guarantee a better performance of the classifiers.

A binary coding scheme is used for the classification process. If there are "p" classifiers to be selected from, then "p" components are reserved for classification selection and they are allowed to have values either 1 or 0. While a 1 at a component represents the presence of the corresponding classifier, 0 represents its absence.

Finally, the combiner selection component is allowed to move through integral values between 1 and the maximum number of combiners present (here 3), with its value directly indicating the combiner to be used by that population member.

For this process, the total data set is divided in two parts: the testing data set and the training data set. For each member of the DHS population, the fitness is calculated by training classifiers (when required, e.g., in RBF, PNN, and GRNN) selected by that vector (decoded from the classifier selection part), finding reduced feature space through feature selection, and finally using these trained classifiers for classification of testing patterns and obtaining decisions from the combiner selected. These decisions are compared with the known classification information for every test data and the average testing accuracy is obtained. This in turn is used as the fitness value for the corresponding vector. In this way, the DHS tunes the parameters, i.e., features selected, classifiers selected, and combiner used, so that the lowest testing error (or highest testing accuracy) is ultimately obtained. The best vector so obtained directly gives the components of the optimal classification machine.

2.3 Results and Discussion

Ten classes of different PQ disturbances are taken for classification and they are as follows—Sag, Swell, Momentary Interruption, Harmonic Distortion, Notch, Sag with Harmonic, Swell with Harmonic, Oscillatory transients, Spike, and Flicker.

The power quality signals corresponding to these ten classes are generated in Matlab [17] using parameterized models with different parameter values. Wavelet transforms of these data samples are then performed to decompose the signals to 13 levels. Equations (2.1)–(2.7) are then used to calculate the features from the decomposed waveform to constitute the feature vector as described previously. Based on the extracted feature, the feature data sets for training and testing are constructed separately. The data set comprises of all seven types of feature vectors for different types of disturbances. The training and testing patterns are chosen in the ratio 4:6 so as to test the real world conditions more effectively where a priori classification information is generally not easily available. The DHS_MD method is used with $m = 13$ and $p = 5$, i.e., the dimensionality of the DHS population being $n = 19$. The optimal machine found is shown in Table 2.1 with complete description about the selected features detailed in Table 2.2. To highlight the performance improvement capability of the proposed method, the overall classification accuracy has been compared with several scenarios in Tables 2.3 and 2.4. While Table 2.3 compares the performance of the machine with a different feature set, Table 2.4 compares that between different classifiers and combiners when fed with the same optimal feature space.

From the comparative study of the power quality classification process presented, it can be clearly seen that the optimal machine performs significantly better in terms of classification accuracy even with a small set of training data patterns.

Table 2.1 Optimally selected features by DHS_MD

Optimal classification machine	
Features selected	4 8 15 40 44 52 55 77 88 90
Classifiers selected	PNN GRNN RBF FKNN
Combiner	Borda count

Table 2.2 Description of optimally selected features by DHS_MD

Serial no.	Selected feature index	Selected feature type
1	4	Energy
2	8	Energy
3	15	Std. deviation
4	40	Kurtosis
5	44	Kurtosis
6	52	Kurtosis
7	55	Mean
8	77	Entropy
9	88	ITD
10	90	ITD

Table 2.3 Comparison of classification accuracy results for different feature set and optimal classifiers-combiners

Features	Classification accuracy
Energy (13)	94.45
Entropy (13)	91.74
Std. deviation (13)	90.46
Kurtosis (13)	94.79
Skewness (13)	89.66
Energy and entropy (26)	91.63
All (91)	92.36
DHS_MD selected features (10)	**99.11**

Table 2.4 Comparison of classification accuracy results for different classifiers-combiners with optimal machine (one optimal set of features used)

Classification accuracy					
	LDA	FKNN	RBF	PNN	GRNN
Without combiner	74.35	80.92	87.99	85.23	84.87
Majority vote (with all classifiers)	89.92				
Borda count (with all classifiers)	94.97				
Bayesian belief (with all classifiers)	93.65				
Optimal classifier-combiner	**99.11**				

2.4 Conclusion

A comprehensive classification method is presented in this chapter that aims to improve the classification process of power quality signals. The discussion of the method has been presented in a general manner that renders it readily amenable to other signal classifications as well. An optimal machine is developed through feature selection, classifier selection amongst few state-of-the-art classifiers, and combiner selection. Further improvements in classification accuracy will come by incorporating more diverse classifiers and combiners into the pool from which the optimal machine can be developed.

References

1. Heydt GT (1991) Electric power quality. Stars in a Circle Publications
2. Daubechies I (1990) The wavelet transform, time/frequency location and signal analysis. IEEE Trans Inform Theory 36:961–1005
3. Mallat SG (1989) A theory of multiresolution signal decomposition: the wavelet representation. IEEE Trans Pattern Anal Mach Intell 11(7):674–693
4. Meyer Y (1992) Wavelets and operators. Cambridge University Press, London
5. Santoso S, Powers EJ, Grady WM, Hofmann P (1996) Power quality assessment via wavelet transform analysis. IEEE Trans Power Deliver 11(2):924–930
6. Gaouda AM, Salama MMA, Sultan MK, Chikhani AY (1999) Power quality detection and classification using wavelet-multiresolution signal decomposition. IEEE Trans Power Deliver 14(4):1469–1476
7. Santoso S, Powers EJ, Grady W, Parsons A (1997) Power quality disturbance waveform recognition using wavelet-based neural classifier, Part 1: theoretical foundation. The 1997 IEEE/PES Winter Meeting, New York, NY
8. Gaing ZL (2004) Wavelet-based neural network for power disturbance recognition and classification. IEEE Trans Power Deliver 19(4):1560–1568
9. Specht DF (1990) Probabilistic neural networks. Neural Netw 3(1):109–118
10. Kuncheva LI (2004) Combining pattern classifiers. Wiley, New York. ISBN 978-0-471-21078-8
11. Chakraborty P, Roy GG, Das S, Abraham A (2009) An improved harmony search algorithm with differential mutation operator. Fundam Inform 95:1–26
12. Storn R, Price KV (1995) Differential evolution – a simple and efficient adaptive scheme for global optimization over continuous spaces. Technical Report TR-95-012, ICSI, http://http.icsi.berkeley.edu/~storn/litera.html
13. Storn R, Price KV, Lampinen J (2005) Differential evolution – a practical approach to global optimization. Springer, Berlin
14. Geem ZW, Kim JH, Loganathan GV (2002) A new heuristic optimization algorithm: harmony search. Simulation 76(2):60–68
15. Uyar M, Yildirim S, Gencoglu MT (2009) An expert system based on S-transform and neural network for automatic classification of power quality disturbances. Expert Syst Appl 36:5962–5975
16. Deokar SA, Waghmare LM (2014) Integrated DWT–FFT approach for detection and classification of power quality disturbances. Electr Power Energ Syst 61:594–605
17. MATLAB, Math Works, Inc., Natick, MA, USA, 2000

Chapter 3
Synchrophasor Measurements

Naim Logic

3.1 Introduction

The introduction of PMUs based on GPS technology has made it possible to obtain synchronized measurements of important power systems quantities. This can be used to get better and more reliable information about the operating status of the power system in real time. The ability to monitor grid conditions and receive automated alerts in real-time is essential for ensuring power system reliability. Synchrophasor technology improves such capability. Synchronized PMUs take sub-second readings system-wide and through visualization and advanced applications provide an accurate picture of grid conditions. With synchrophasor technology, operators can monitor reliability metrics, grid dynamics, identify and diagnose system problems, system stresses, oscillations, and other abnormal situations that may occur in the power systems to enable them to take proactive actions to prevent blackouts, reduce the footprint of blackouts, and enable faster recovery after events. Given today's pressures on grid operators to manage an ever increasingly complex grid, PMUs may hold the key to reliable operations: grid monitoring and visualization, decision support, and post-event assessment. In addition, this technology has the potential to improve efficiency by allowing operation closer to inherent thermal physical limits with equal or greater safety margins and by increasing existing transmission lines use based on dynamic ratings and margin assessment.

A growing community of researchers and utility experts are working on practical applications and installations of this technology around the globe. An increasing number of transmission system operators world-wide are evaluating the benefits of this technology and implementing demonstration projects. Synchrophasor

N. Logic (✉)
Computer Application Department, System Operations Applications Group,
Salt River Project, Phoenix, AZ, USA
e-mail: naim.logic@srpnet.com

E. Kyriakides et al. (eds.), *Electric Power Engineering Research and Education*,
Power Electronics and Power Systems, DOI 10.1007/978-3-319-17190-6_3

technologies establish the foundation needed to operate and control the future power grid as it becomes more complex with increasing reliance on renewable energy generation, continued growth in electric transmission, and greater diversity of end-use electrical loads such as sophisticated power electronics.

This chapter describes synchrophasor technology features giving more details about research projects where Prof. Gerald T. Heydt was helping in solving power system industry issues related to the usage of synchrophasor measurements, ranging from the integrated calibration of synchronized phasor measurements through the identification of synchronous generator dynamic parameters.

3.2 Background

Leveraging advances in computing, high-speed communications and graphics synchrophasor technology will benefit power system control room operation by improving intuitive and informative visualizations, better grid condition situational awareness, smart and integrated alarms and alerts, sophisticated decision support tools, automate and distribute some less important activities to let operators work on high-value activities and challenges.

PMUs provide voltage and electric current measurements along with frequency and rate of change of frequency. This data can be used to provide early detection to prevent grid disturbance events, assess and maintain system stability following a destabilizing event, as well as alerting system operators to view precise real-time data. This capability reduces the likelihood of an event causing widespread grid instability. In addition, the use of phasor measurements is very valuable for postmortem event analysis to understand the cause and impact of system disturbances. In addition, PMU data is improving power system fault location detection and analysis.

Phasor data is also useful in validating the dynamic models of generation resources, energy storage resources, and system loads for use in transmission planning programs and operations analysis, such as dynamic stability and voltage stability assessment. This technology will have an important role in determining dynamic system ratings and allow for more reliable deliveries of energy, especially from renewable generation location to load centers. In addition, the higher penetration of renewable generation changes power system dynamic characteristics making the application of synchrophasor technology more desirable. More renewable generation resources driven by power electronic devices decrease overall power system inertia causing speeding up of possible and inevitable grid events. The existing Supervisory Control and Data Acquisition (SCADA) system is not able to recognize such events in a timely manner. Only PMUs and other smart devices are able to do that since they are about 100 times faster.

PMUs were developed from the invention of the Symmetrical Component Distance Relay (SCDR). The SCDR development outcome was a recursive algorithm for calculating symmetrical components of voltage and current [1]. Synchronization is made possible with the advent of the GPS satellite system [2]. The PMU records the sequence currents and voltages and/or their phase values and time stamps the reading with time obtained by the GPS receiver. It is possible to achieve accuracy of synchronization of 1 μs or 0.021° for a 60 Hz signal. A growing number of power system protective digital relays are being introduced to the market with the ability to be used as phasor measurement units in addition to their protective relaying function.

The IEEE has recognized the need for standards for PMUs. The first standard for PMUs, IEEE 1344 [3], was completed in 1995, and reaffirmed in 2001. In 2005, it was replaced by IEEE C37.118-2005 [4], which was a complete revision and dealt with issues concerning the use of PMUs in electric power systems. The specification describes standards for measurement, the method of quantifying the measurements, testing and certification requirements for verifying accuracy, and data transmission format and protocol for real-time data communication. This standard was not comprehensive since it did not attempt to address all factors that PMUs can detect in power system dynamic activity. A new version of the standard was released in December 2011, which split the IEEE C37.118-2005 standard into two parts: C37.118-1 dealing with the phasor estimation [5] and C37.118-2 the communications protocol [6]. It also introduced two classifications of PMU, M—measurement and P—protection. M class is close in performance requirements to that in the original 2005 standard, primarily for steady state measurements. The P class has relaxed some performance requirements and is intended to capture dynamic system behavior.

Because PMUs have the ability to directly measure phase angles at high sampling rates and accuracies, they are prompting a revolution in power system operations as next generation measuring devices. Figure 3.1 shows lack of

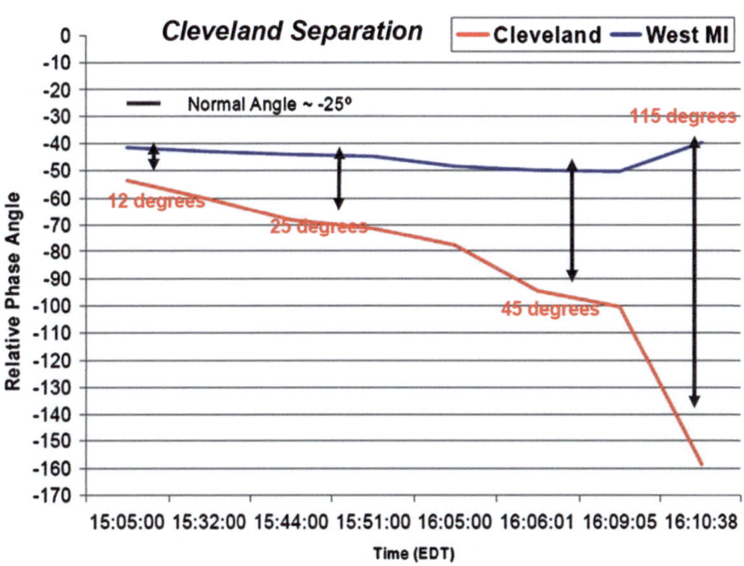

Fig. 3.1 August 14th, 2003 eastern interconnection blackout [7]

wide-area visibility during the August 14th, 2003 Eastern interconnection blackout. The disturbance in the power system usually develops gradually—from several minutes to milliseconds, depending on the type of the disruptive event. In this case, the increasing phase angle difference across the grid for 64 min was an indicator of increasing system stress. Unfortunately, this information was not available at the time of the event. The figure is based on data from forensic blackout analysis. The blackout was estimated to cost the US economy $6–$8 billion.

PMUs typically sample grid conditions at a rate of several hundred times per second and use this sampled data to calculate phasor values for electric voltage and current, at a rate of 30 or more per second. A phasor is a complex number that represents the magnitude and phase angle of the sinusoidal waveforms of voltage or current at a specific point in time. In North America PMUs are streaming 30, 60, or 120 samples per second, having a rate of 30 samples/s as an industry standard. A Phasor Data Concentrator (PDC) is collecting phasor measurements from multiple PMUs aligning these signals according to time tags based on the GPS signal. The PDC receives many data streams from different PMUs, packages them together and broadcasts the packets as a user datagram protocol in a PDC stream format. The IEEE C37.118 protocol has been designed to connect directly with the PDC output, read the phasor data stream in real-time, and calculate scaled and derived values such as Megawatt (MW) and Megavolt-ampere reactive (MVAr). Once received, the raw data is processed to remove erroneous data and filtered for noise to improve data quality. Additionally, for the visualization tool, it is necessary to set filtering options, enter PMU location longitude and latitude, and define alarm and event alarming attributes. The parsed phasor data received from the PDC and derived quantities are stored in a real-time memory buffer. Additionally, the data is usually stored in a database for long-term trending and reporting purposes, as well as for forensic analysis.

Time stamping allows measurements taken by PMUs in different locations and by different transmission operators to be correlated and time-aligned, and then combined accurately. Such data can provide a comprehensive picture of transmission system operations across an entire transmission region or interconnection.

3.2.1 Infrastructure and PMU Placement

One challenge related to the usage of synchrophasor technology is the communication infrastructure, which lacks the bandwidth to handle the data traffic produced by the smart devices, needs enhanced security and it must maintain a high degree of reliability if the data is used for control decisions. Initially, synchrophasor hardware infrastructure was designed as a research and development pilot project. As utilities and power industry have become more aware of PMU benefits, along with government incentives to deploy more units, it has become clear over the years that developing a production-quality synchrophasor architecture system is needed to support grid reliability and renewable integration. Therefore, a high-speed communications network and secure infrastructure is essential for synchrophasor technology implementation. With all of these comes the need for enhanced cyber security plans, comprehensive monitoring of communications traffic, and strategies to ensure reliable and secure data transfer.

Currently, almost 1,700 PMUs are deployed throughout North America. PMU placement is often difficult to determine. Ideally, utilities would like to have PMUs installed at every bus in their system but, in reality, that is not economically feasible. Therefore, future PMU installations need to be prioritized by taking into account the data requirements of all the different synchrophasor applications. These requirements are resulting in the suggestion to install PMUs in high voltage substations, large generation power plants and load centers, major transmission lines, substations with remedial action schemes, renewable generation plants, etc.

3.2.2 Data Quality and Management

With the ongoing investments in a smarter electric grid, new algorithms and devices are being developed. Synchrophasor applications for electric transmission systems are one of the most critical smart grid technologies. The high quality of synchrophasor measurements is vital for most of these applications, especially for real-time control where achieving an accurate picture of the current grid state is essential.

With the synchrophasor applications transitioning from the pilot project to production and the increase in data availability it is necessary to develop a strategy for data retention and storage. Also, it is very important to make this data more accessible. Currently, displaying the data in raw or processed form require custom applications. However, to maximize the benefit from this technology, it is necessary to make this data readily available to all utility systems that can use it.

3.2.3 Advanced Applications

Another major challenge is the lack of the available applications that assimilate and provide meaningful, understandable visual displays of the extensive data produced by the smart devices.

Unprecedented volumes of data coming into the control room require efficient computing techniques and application tools to be able to extract maximum benefits from this promising technology. Precursors to the actual power system disturbance are often only found in synchrophasor data. Advanced applications software providing these capabilities are necessary to realize the full potential of synchrophasor technologies. They will improve grid reliability, power quality, asset utilization, and efficiency in grid planning and operations. These are the ultimate benefits of synchrophasor technologies.

To address fast-occurring grid events and free up grid operator attention for high-priority decisions synchrophasor technology is providing better wide-area situational awareness and visualization tools, better state estimation, faster contingency analysis, better transmission pathway and congestion management, pattern recognition, and libraries of past grid events enabling faster event identification and appropriate reaction. More automated actions out on the grid with synchrophasor-driven decision-making and switching, including voltage management at generators, automated reclosing decisions informed by phase angles, and dynamic stability management. Synchrophasor technology can change operators' roles and tasks with usage of dynamic nomograms, faster power system restoration, system inertia monitoring, developing linear state estimation, detecting imminent cascading, and introducing adaptive protection.

Synchrophasor technology is introducing the key attributes of smart grid at power system transmission level: interactive with consumers and markets, optimized to make best use of resources and equipment, predictive rather than reactive in order to prevent emergencies, integrated by merging monitoring, control, protection, maintenance, marketing, … distributed across geographical and organizational boundaries, self-healing and adaptive nature, more secure from attack, etc. Recent surveys on PMUs and their applications in power systems can be found in ref. [8].

3.2.3.1 Visualization

Synchrophasor technology is improving power system monitoring and visualization to aid power system operators' situational awareness and help them forestall grid collapse through better recognition and response to evolving grid events. PMUs are deployed across an area as wide as an entire interconnection, PDCs are collecting data and display it for operators to understand grid conditions indicating possible levels of stress in the grid, such as areas of low voltage, frequency oscillations, or rapidly changing phase angles between two locations on the grid. Many applications have diagnostic capabilities that can identify grid stress (measured by the changing phase angles of synchrophasors at different substation locations, phase angle separation), grid robustness in terms of system events (oscillations, damping, and trends), instability (frequency and voltage instability), or reliability margin (which describes how close the system is to the edge of its stability boundary) providing appropriate graphics and visualizations, basic data archiving, the ability to drill down into specific locations or conditions on the grid, and playback capabilities.

3.2.3.2 Frequency Stability Monitoring

PMUs measure power system frequency, which is a key indicator of the balance between generation and load in the power system. Abrupt changes in frequency due to major losses in generation or load can compromise power system stability and lead to a blackout.

3.2.3.3 Voltage Stability Monitoring

Synchrophasor systems can be used to monitor, predict, and manage the voltage on the transmission system of the power grid. Many transmission systems are voltage stability-limited, which means that the voltage cannot exceed a certain level without causing system stability problems (instead of thermally limited). Voltage collapse can happen very quickly if these voltage stability limits are reached or exceeded. The high resolution of synchrophasor measurements provides the ability to map changes in voltages at a bus to power flow changes on connected lines, which measures voltage sensitivity at that point. High voltage sensitivities could be an indicator of possible voltage stability problems. The phasor measurement based voltage stability analysis application assesses the power (or current)-to-voltage system operating point and sensitivities at a sub-second resolution. The calculated adaptive voltage stability margin can be expressed as active and reactive power. This will provide system operators not only the power transfer limit to a load center, in terms of active power, but also the reactive power support needed at this load center.

3.2.3.4 Phase Angle Monitoring

The phase angle difference between buses measured by PMUs on the transmission grid is an indication of system stability and system stress. An angle difference within a predetermined limit is acceptable but needs to be monitored closely for early warnings. An increasing margin of phase angle difference can be a serious problem when the deviation gets large enough to cause voltage and system instability. System operators can be assisted by monitoring angle separation or rate-of-change of angle separation between two buses or two parts of a grid to determine stress on the system. One application for synchrophasor based situational awareness and trending tools is to have them show the trend in phase angles compared to phase angle limits in order to warn operators when the stress is increasing. Such a tool offers intelligence to the power system operator. When phase angles exceed critical limits, operators can perform corrective actions.

In some cases separation is unavoidable in a power grid due to inter-area oscillations, out-of-step protection, or cascading. PMUs can help monitor inter-area oscillations and predict when a controlled system separation is needed. A PMU based controlled separation scheme has been developed to address the key issues: where to separate, when to separate, and how to separate? Another important application of phase angle monitoring is during restoration. The phase angle value across an opened tie line or an opened circuit breaker would guide an operator in circuit breaker closing.

3.2.3.5 Oscillation Detection and Analysis

Detecting power system oscillations requires high resolution data, which is available through synchrophasor technology with sampling rates of 30, 60, or 120 samples per second. This further enhances operator situational awareness. Oscillations are not observable using SCADA technology, with sampling rates of one sample every 2–4 s. It is every operator's objective to keep system oscillations under control while maximizing power transfers and maintaining system reliability. Monitoring devices such as PMUs can play a critical role in dynamic conditions.

System oscillations are expected in a large interconnected network of loads and generators. During normal operations, generators have the ability to retain synchronization speed with other interconnected generators and can maintain system stability. If the system is properly designed and operated and a disturbance occurs, the oscillations are damped and the system naturally returns to equilibrium quickly. However, if the disturbance is large or the interconnected system is somewhat weak, the oscillations caused by disturbance could grow causing the system to become unstable and break apart leaving large areas without power.

In some cases when system damping is inadequate, a disturbance on the grid, such as scheduled or unplanned transmission line outages can cause adverse and serious system stability problems. Sometimes only a slight variation in system operating conditions might exist in which an oscillatory mode becomes lightly damped. Under such circumstances, it is possible that the system operators fail to notice this insecure state and miss the opportunity to remedy the situation before it is exacerbated by a system disturbance. PMU data can be used to compute the damping ratio and determine the magnitude and energy associated with oscillations with certain frequency (so-called oscillation modes) in real-time operations. A wide-area damping control system has recently been developed to enhance the damping performance of inter-area oscillations.

Variable generation resources can strain grid stability and as such, system oscillations are becoming a much more serious treat. Monitoring, detecting and identifying low damping oscillations in a system with renewable resources is essential for system operators since this allows them to start control measures to damp out the oscillations by re-dispatching or forcing reduction in power generation. When oscillations are detected, a control signal can be generated and sent to the offending generator's excitation system to regulate its voltage and bring it into synchronization or reduce its output to a level that is no longer threat to the system.

Data obtained from PMUs can help identify the causes of some of the stability issues in the system. For example, some of the unstable modes are local where a small group of generators is out of synchronization with the rest of the generators in the system or inter-area where many generators in one part of the system are out of synchronization with the rest of the system. There are also some higher frequency oscillation modes that are caused by poorly tuned exciters, power system stabilizers, governors, or static VAR compensators.

Off-line and post-disturbance analyses are very useful in understanding grid dynamics. Using the data recorded by PMUs is vital in the calculation of the distinctive frequency responses contributed by each generator. Off-line analyses, such as modal analyses (using event or ambient data) can identify oscillation and damping modes.

3.2.3.6 Dynamic Capability Rating

Dynamic Capability Rating Systems on transmission circuits continuously monitor ambient conditions, such as line tension, temperature, or wind speed, and allow lines to be reliably loaded closer to their true operational capacity. Often this means they can carry electricity at higher levels than nominal limits. However, in some conditions, they can warn operators of situations where the capacity of the line is reduced. PMUs can monitor the precise grid synchrophasor measurements and are typically installed at substations or at power plants, at a variety of voltage levels. Depending on location and surrounding network configuration, a PMU can be used to monitor transmission lines, transformers, and/or generators.

Congestion management is a critical function performed by grid operators in real time and power schedulers in the advance market. It involves generation dispatch (in day-ahead markets) and re-dispatch (in real-time markets) to satisfy the demand in an economic manner without violating the transmission limits. The traditional approach to real-time congestion management compares actual flow on a line or path against a Nominal Transfer Capability (NTC) that is calculated in advance using an off-line methodology—seasonal summer or winter ratings of lines are typically set based on fixed assumptions regarding ambient temperature, wind speed, and solar heating input to arrive at a conservative figure for transmission line conductor capacity based on thermal limitations, voltage limitations, or stability limitations; whichever happens to be the most restrictive in any given situation. Since the assumptions used in off-line NTC calculations are often conservative and result in excessive margins in the congestion management process, this may lead to unused transfer capability and lost opportunity costs in the dispatch process. PMU measurements may be used to calculate more accurate path limits in real-time, thus providing more "room" to manage congested lines and paths. Higher scan rate and more precise PMU data allow rapid computation of Real-time Transfer Capability (RTC) applied to the critical stability-limited and voltage-limited paths. In many cases such RTCs will exceed their respective seasonal NTC and reduce the need for real-time congestion curtailments.

3.2.3.7 Resource Integration

PMUs are expected to be particularly useful for improved monitoring, managing, and integrating of distributed generation and renewable energy resources. One of the challenges in integrating these resources is how to identify and respond to their power generation variability. In a conventional system, frequency is controlled by large central rotating generators. However, as more renewables come online, they challenge the ability of the power system to control the system frequency because, with renewables with much less inertia, it can change much faster than in a conventional power system without renewables. This variability alters the frequency behavior of the interconnected system and could adversely impact the grid's stability performance. Real-time monitoring of frequency behavior enables operators to take appropriate actions to maintain stability.

3.2.3.8 State Estimator

State Estimation (SE) is an essential component of a modern Energy Management System (EMS). The performance of other application programs depends on the accuracy of data provided by the SE. The SE processes a set of real-time redundant measurements. Measurements received at the control center include line power flows, bus voltage and line current magnitudes, generator outputs, loads, circuit breaker and switch status, transformer tap positions, and switchable capacitor bank values. The SE solution will provide an optimal estimate of the system state based on the available measurements and on the assumed system model. This solution consists of voltage phasors (magnitudes and angles) for all buses in the system model. Until recently it was nearly impossible to measure voltage and current phase angles, but synchrophasor technologies now allow one to calculate these phase angles. So PMUs are able to measure phase angles and it was natural to consider synchrophasor technology benefits as SE improvements and accuracy enhancement.

3.2.3.9 Dynamic Model Validation

Dynamic modeling and validation of complex power system components are essential for optimal operation, safety and security of the grid. Also, important decisions on new capital investments are dependent on the accurate modeling of the different components in the system. However, due to a large number of components in the power system and time-varying characteristics, acquiring the precise model and parameters is not a trivial task. The unpredictable nature of system disturbances and modeling shortcomings to reflect real-time dynamic response of the system during events add complexity into finding a true system representation.

In addition, the rapid increase in renewable resources penetration in the system further highlights the need for more accurate models and improved validation methods. The new models need to give an accurate representation of the aggregated nature of renewable resources, such as solar and wind. System planning for such systems can be difficult without an accurate comprehensive model. Failure of a model to predict or replicate the system's response to a disturbance is an indication of model weakness and inaccuracy.

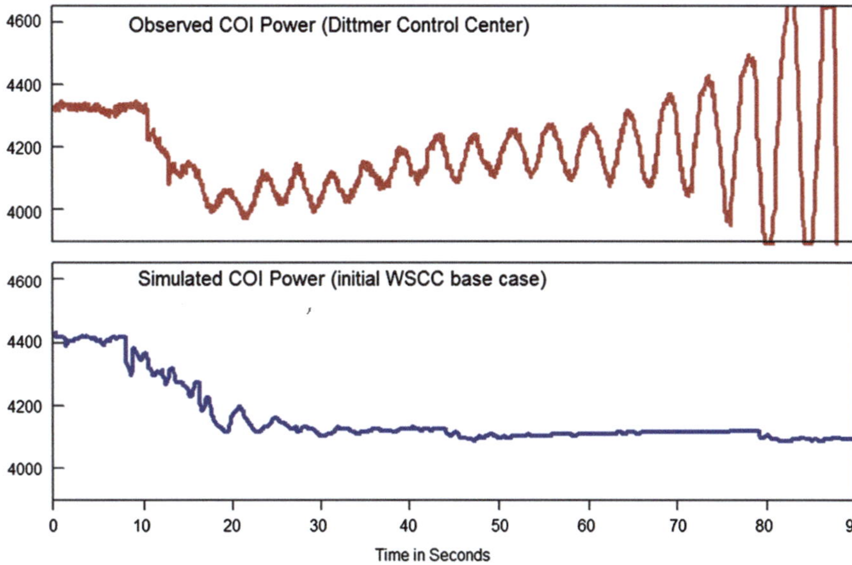

Fig. 3.2 August 10th, 1996 WECC system separation into four electrical islands [9]

Figure 3.2 shows the discrepancy between the simulated California-Oregon Intertie (COI) power transfer and the active power flow recorded at the Malin substation on August 10th, 1996 when the WECC system separated into four electrical islands. The predicted power flow on the COI was not able to capture the growing oscillation phenomena. During off-line studies, the simulated COI power response resulted in a stable response without significant oscillations while in reality the system presented an unstable response with undamped and growing oscillations.

Power system model validation by field test of any single power system component can be time consuming, expensive, and even sometimes an unachievable goal. An easier way to validate system models is to use data from real event recordings, compare them against the simulated dynamic response and then make the necessary corrections to the models. In addition, improving dynamic models will provide more accurate results when performing system operation limits studies which might relax or tighten operational limits resulting in more reliable and safe system operations.

One of the useful applications of PMUs is the validation of the load models for advanced voltage stability assessment. Accurate modeling of the aggregated loads is essential for that application, as load characteristics critically influence the system voltage behavior.

A major problem faced by the power system operators during the disturbance is load recovery, which increases the reactive power consumption and leads to reduced voltages in the transmission system. In the US Southwest, with a lot of single phase induction motors, a known issue is Fault Induced Delayed Voltage Recovery (FIDVR). It is thus very important to have realistic load models (both static and dynamic).

3.2.3.10 Forensic Analysis

Post-event analysis is necessary to ensure that lessons are learned to correct problems that previously led to an event, to train system operators on the lessons learned, and finally to take measures to correct the problem. Synchrophasors are essential for post-event analysis of power systems. Data synchronization is critical for the sequence of event reconstruction, particularly for complex events where the switching of many devices in the system occurred in a short-time frame. Prior to synchrophasors, it could take many months of investigation to reconstruct the sequence of events that caused a blackout. However, having synchrophasors in place greatly reduces the time required to complete a post-event analysis to days or hours.

3.3 State Estimation Improvements

In the early stage of discovering benefits of synchrophasor technology the focus of research efforts was on how the new technology of phasor measurement units can be used to enhance state estimation in electric power systems. There is a school of thought that the measurements from the PMU are far superior of SCADA data used in traditional state estimation and should be collected and used separately from this data [10]. Others admit that there is difference in the information and it is viable to use PMU measurements with SCADA data [11]. A dramatic improvement in the state estimate has been seen by using a three-phase model and the use of GPS synchronized measurements [12].

In one approach to enhance SE accuracy the sensitivity analysis of three condition indicators to added state measurements was performed:

- The condition number K_{G} of G, using the 2-norm, where:
  - G is the gain matrix $G = H^t H$.
  - H is the process matrix in SE approach.
- The “distance” from G to the nearest singular matrix S as d.
  - $d = \frac{\min}{S} ||H^t H - S||_2$, where the minimum is taken over all possible singular matrices S.
- The scaling factor $F = \mathrm{d}K_G$

PMU placement can be performed using several different criteria including security concerns, observability, and improvement in state estimation resulting in usage of eigenvectors—the use of eigenvector of the smallest eigenvalue of H [13].

Condition indicators were studied in order to supplement measurement placement algorithms. The analysis also shows that the concept of using condition indicators for measurement placement is consistent with redundancy analysis. Eigenvalue sensitivity analysis is used to improve the condition indicators [14].

Prof. Heydt also made an interesting approach to a distributed state estimation algorithm suitable for large-scale power systems. In the decomposed subsystems, the placement of synchronized phasor measurements was investigated. They were applied to aggregate the voltage phase angles of each decomposed subsystem in the distributed state estimation solution obtained by using a sensitivity analysis based update at chosen boundary buses [15].

Recently, Prof. Heydt has incorporated synchronized phasor measurements into a linearized, three-phase, distribution class state estimation algorithm for applications in smart distribution systems [16]. Knowledge of bus voltage phase angles from synchronized phasor measurements improves the state estimation process for power systems, and is shown for distribution circuits having the advantage of providing full three phase detail.

3.4 The Time Skew Problem in PMU Measurements

The accuracy of PMU measurements is inherently higher than the accuracy of the conventional SCADA measurements. One of the important features of PMU measurements is the synchronization indicated by the time stamps. The GPS provides the one pulse per second (1 PPS) signal used for the same time sampling at different locations. The precision of the 1 PPS signal is within 1 μs.

The faulty synchronization of PMU measurements is termed as a "time skew" problem. Time skew exists between measurements from different PMUs even though they are tagged with the same time stamps. The origin of the time skew is due to the inaccuracy of the sampling clock associated with specific PMUs. A resynchronization of all sampling clocks is achieved every second due to the 1 PPS signal. This time skew problem is identified, but solving this problem is not an easy task and might not be accomplished in a short term due to procedural constraints in the field. The recognition of the time skew problem is of great importance to applications of PMU measurements in power system monitoring, protection and control.

All PMUs are receiving the 1 PPS from the GPS system, and sampling clocks are phase-locked to this time pulse. However, the sampling clocks might not be exactly synchronized during each one second interval due to the different accuracy of the clocks, and time skew originates between measurements from different PMUs. Two examples with less accurate clocks are shown in Figs. 3.3 and 3.4.

The question arises as to how to judge whether a clock is accurate or not. The voltage phase angle change is used to accomplish the goal. The examples shown in Fig. 3.5 from actual measurements illustrate this point. The curve in Fig. 3.5a depicts the voltage phase angle change based on measurements from a PMU associated with a more accurate clock. It is observed that the voltage angle changes are relatively small numbers, and the absolute values of the angle changes are in a range of [0, 0.06] degree. As for the curve in Fig. 3.5b, this plot depicts the voltage phase angle change based on measurements from a PMU associated with a less

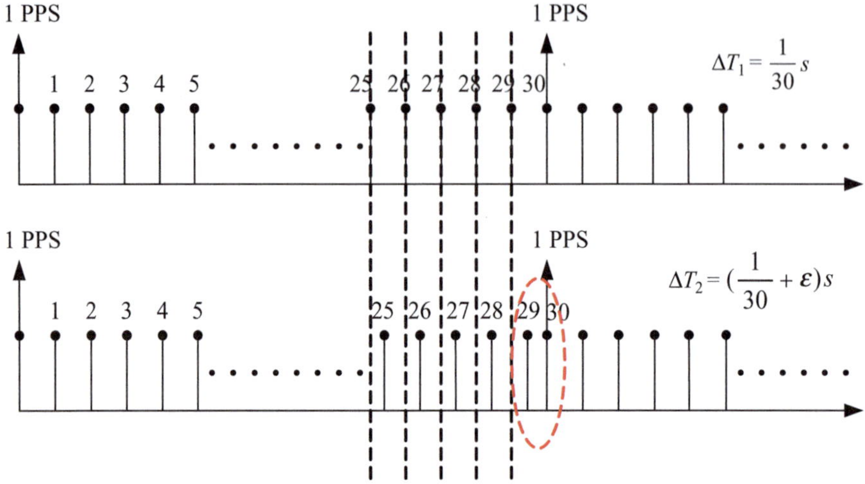

Fig. 3.3 Time skew originating from a less accurate clock when ΔT_2 is slightly greater than ΔT_1 [17]

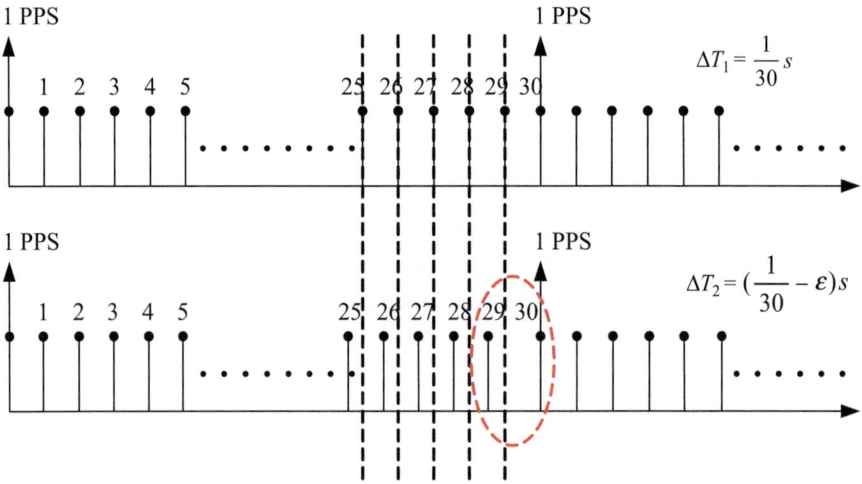

Fig. 3.4 Time skew originating from a less accurate clock when ΔT_2 is slightly less than ΔT_1 [17]

accurate clock. It is clearly seen that spikes occur every second corresponding to the large changes as stated in the previous discussion. The absolute values of the spikes are about 0.2°. It should be noted that the resynchronization takes place at one reporting step after the 1 PPS signal shown by the time stamp.

To compensate the time skew error, a Kalman filter model is used. The Kalman filter is a least squares error estimator using measurements and system information to model the process dynamically [18]. It is a recursive optimal estimator based on

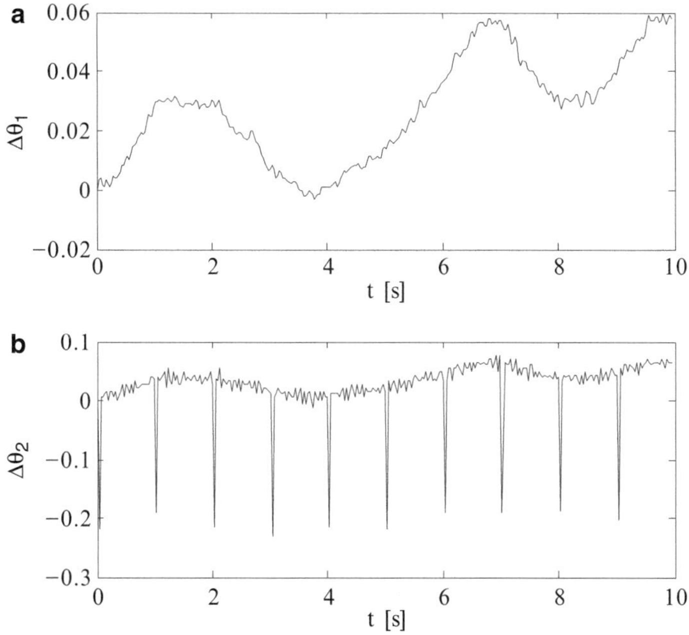

Fig. 3.5 Voltage angle changes from two PMUs in a time period of 10 s (**a**) PMU with an accurate clock and (**b**) PMU with a less accurate clock [17]

the state space representation where the equations that describe the Kalman filter estimator are divided into two steps: time update (predictor) equations and measurement update (corrector) equations. In essence, the Kalman filter is used as a dynamic state estimator. Analysis is performed to answer two questions: ascertain whether the error due to the time skew problem is constant or not, and determine the error identified in the first answer.

The raw PMU relative voltage phase angle measurements are shown by the curve in Fig. 3.6a, and the processed measurements are shown by the curve in Fig. 3.6b in a time period of 3 min. It is easy to notice that the raw PMU relative voltage phase angle measurements vary in range of (−1.5, −1.2) degrees and the processed measurements vary in a much smaller interval, which reflect the actual system operation condition.

Analysis shows that the time skew error for one specific PMU with time skew problem could be treated as a constant. Then during each one second interval, the time skew errors are compensated before PMU measurements are used in further power system applications. Since the computational cost is low, this method is easy to be applied on practical PMU measurements [17].

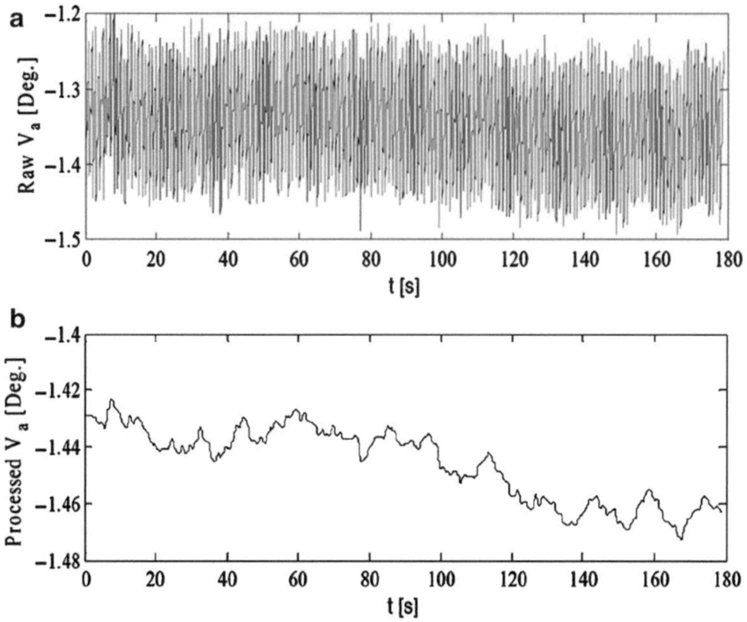

Fig. 3.6 Relative voltage phase angle measurements (**a**) without and (**b**) with compensation

3.5 The Integrated Calibration of Synchronized Phasor Measurement Data

Perhaps the most significant breakthrough in recent years in power system instrumentation has been in the development of synchrophasor measurements via PMUs. This research work was concerned with the error, accuracy, and validation of PMU measurements from a practical electric power system.

Measurements have become a key element of power system operation and they are instrumented mainly utilizing potential and current transformers (PTs and CTs) and voltage, current, and power transducers. A commonly used type of voltage transducer is a Capacitively Coupled Voltage Transformer (CCVT). The IEEE Standard 57.13-2008 [19] prescribes the accuracy (maximum error) of relaying class CTs as shown in Table 3.1. As an indication of ratio and phase angle accuracy, Table 3.2 shows the permissible error for PTs in ref. [19] and CCVTs in ref. [20].

The PMU measurements are a time sequence data series. There is noise and possibility of bad data in the measurements. Data filtering is applied to remove the corrupted data points from the PMU measurements. The objective is to obtain a more accurate estimate of the system states based on PMU measurements.

Table 3.1 Accuracy ratings for relaying class CTs [19]

Limits of ratio error relay class	At rated current (%)	At 20 times rated current (%)
C and T classification	3	10
X classification	1	User defined

Table 3.2 Permissible error for power system instrument

Application	Maximum error in ratio	Maximum error in phase
PTs [19]		
Revenue metering	±0.1 %	±0.9 mrad (3 min)
Other applications	±1.2 %	±17.5 mrad (1°)
CCVTs [20]		
Relaying	<1.2 %	±63 min
Metering	0.3–1.2 % depending on class	±16 min to ±63 min depending on class

Even though the power system is never truly static, the power system states vary little when there is no large disturbance in the system. In this case, a Kalman filter can suitably filter measurements from the PMUs. There are two conditions that need to be considered: one deals with bad data detection, and the other relates to the system state change detection. When a bad measurement is detected, the measurement should be replaced by an estimate of the system state or be simply discarded. When a system state change is detected, the measurement should be retained. The question arises as to how to determine whether the "abnormal" data observed are bad data or true system changes. When the redundancy of PMU measurements is available, this procedure is applied as a key step in adaptive Kalman filtering of PMU measurements to overcome the insensitivity of a conventional Kalman filter.

In addition, another methodology of identifying bad data from PMU measurements based on Power Transfer Distribution Factor (PTDF) is considered. The definition of PTDF is the relative active power change through a particular branch due to the change in active power injections. Providing that the number of PMUs installed in the network is sufficient to satisfy the redundancy requirement, this procedure could be conveniently implemented in online analysis of PMU measurements. It helps to determine whether the system state changes reflected by PMU measurements are physical changes or bad measurements, and can be applied in the adaptive Kalman filter approach.

Moreover, a novel method for bad data detection and identification of PMU measurements based on voting criteria is considered. Under certain PMU installation scenarios, the topological relationship in the system could provide additional information that is helpful for bad data detection. The method is especially beneficial for utilities with a relay based PMU network. This method could be applied under the condition that the redundancy requirement of PMU measurements is not satisfied to implement the bad data detection and identification procedure mentioned previously. Basically, this is a preprocessing method to detect and identify bad measurements from PMUs which are closely related to the physical topology.

This research deals with the error, accuracy and validation of PMU measurements from a practical electric power system. Data from operations planning cases are integrated with measurements taken during actual operation. The basic concept of integrating all available information to enhance power transmission operational measurements is exemplified by the inclusion of operations planning data into the measurement process. In order to utilize PMU measurements in system control, validation of the data from PMUs is required. The calculated Transmission Line (TL) impedance based on the raw measurements from PMUs shows a difference from TL parameters from planning data indicating that calibration is needed to improve the accuracy of PMU measurements. Calibration or Correction Factors (CFs) are introduced to calibrate each of the electrical current and voltage phasors measured. There could be errors both in magnitude and angle of synchrophasors. Three different cases presented in Figs. 3.7, 3.8, and 3.9 are considered to calculate the CFs.

These calculated CFs are complex quantities indicating correction of both magnitude and phase angle of the synchrophasor measurements and they are

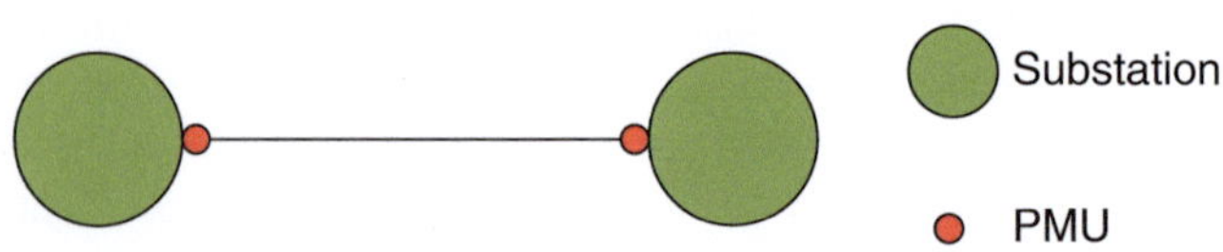

Fig. 3.7 Transmission line with two PMUs [21]

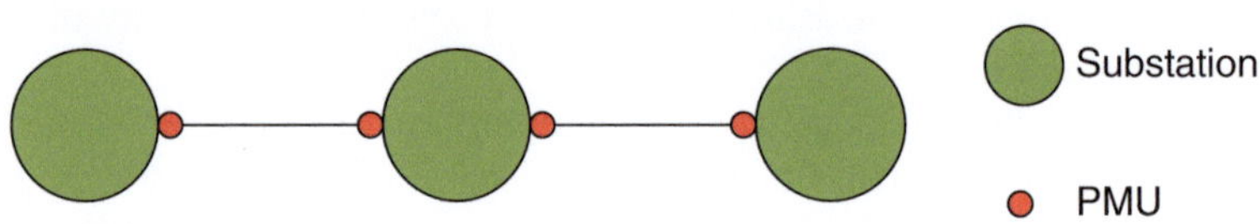

Fig. 3.8 Two transmission lines with four PMUs [21]

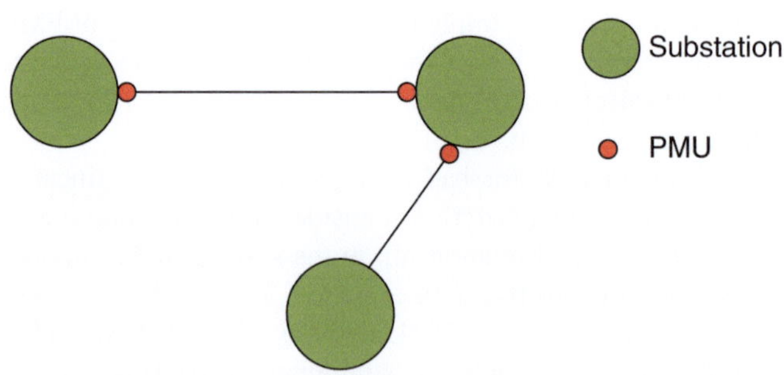

Fig. 3.9 Two transmission lines, with one PMU each [21]

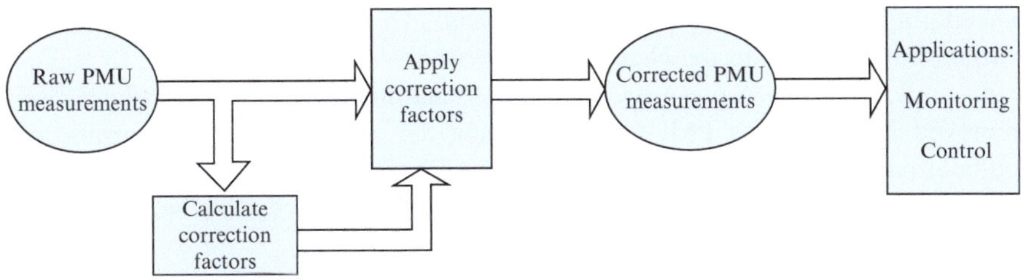

Fig. 3.10 Implementation of correction factors methodology [21]

introduced to make the measurements and TL parameters 'match'. The correction factors are utilized to calibrate the PMU measurements. Comparison of the active and reactive power mismatches before and after calibration is used to determine the effectiveness of the correction factors.

Figure 3.10 depicts the implementation of the proposed correction factors for PMU measurements. A data processing algorithm is implemented and used as preprocessing before raw PMU measurements are applied for power system monitoring and control. By analyzing the characteristics of CFs, a novel calibration process of PMU measurements is proposed which is applicable in real-time power system operation [21].

3.6 Impact of PMU Measurement Buffer Length on State Estimation

Attention is turned to an important specific issue when integrating PMU measurements into state estimation. In practical systems, the reporting rate of PMU measurements is much higher than the conventional measurements collected by the SCADA system. If the reporting rate of synchrophasor data to SE is one second, which is currently the industry standard, the question is which measurement inside the one second time interval to use for SE? The first, the last, the middle, the mean, or something else?

If a buffer of PMU measurements is utilized (i.e., a memory buffer of length N_{bl}), the variance of the PMU measurements could be calculated directly. As for the noise in the PMU measurements (which is generally assumed to be normally distributed with zero mean), this is a quantity that is a random process and needs to be modeled and integrated appropriately. Accordingly, more measurements should be included to yield the best estimate of the "true" value of the state when the system is static. However, in reality, power systems are never truly stationary or static. Again, the more measurements that are included, the larger is the deviation that would result due to system dynamics. These contradictory aspects raise the question of how to choose the buffer length of the PMU measurements in preprocessing data used in the state estimation. The first aspect is referred to as uncertainty due to the noise in the measurements. The second aspect is referred to as variation of data due to the change in the system states. It is important to make a good tradeoff between these two aspects. Assuming that the bad data have been removed from the raw PMU measurements, the goal is now to determine the optimal buffer length.

The equation relating the measurements z and the state vector x is:

$$z = h(x) + e$$

where z is the measurement vector, e is the measurement error vector and $h(x)$ are functions of the system states x. The residual vector r is defined as:

$$r = z - H\hat{x}$$

The Euclidean norm (2-norm) of r is:

$$r_{\text{index}} = \|r\|$$

The scalar r_{index} defined above is used as an index to check the impact of different buffer length of PMU measurements on the state estimation. It should be noted that there is no noise added to the conventional measurements. For a certain buffer length N_{bl}, the simulated signal is divided into subsets. The mean value of the measurements in each subset is calculated. Then, each of the mean values is used in the state estimation to obtain r_{index}. In order to obtain the probability distribution function of r_{index}, a Monte Carlo simulation is applied.

If the cumulative distribution of r_{index} is considered when the buffer length varies, the result is that r_{index} decreases as the buffer length increases. When the system is static, the larger the N_{bl} is, the better the state estimation becomes. If only the uncertainty due to noise in the measurements is considered, a large buffer length would yield a better (lower) r_{index} than a small buffer length.

Attention is turned to the design of the optimal input data buffer length. If the variation of PMU measurements due to the change of the system states could be quantified, the optimal buffer length is obtained at the point where the summation of the residuals due to the uncertainty and variation of measurements is minimum. However, it is difficult to separate the impact of the uncertainty and the variation of the measurements. Also, it is difficult to describe the variation of system states in a probabilistic sense. Figure 3.11 depicts the possible changing trends of the system states shown by the dashed curves. It is possible that the system is static, as shown by the dashed straight line, or varying as shown by the other two dashed curves. As explained previously, it is important to insure the best possible tradeoff between reducing noise uncertainty impact on estimation and improving system change tracking accuracy. The first objective implies increasing the buffer length whereas the second one means decreasing it. Thus, the goal consists of finding an optimal

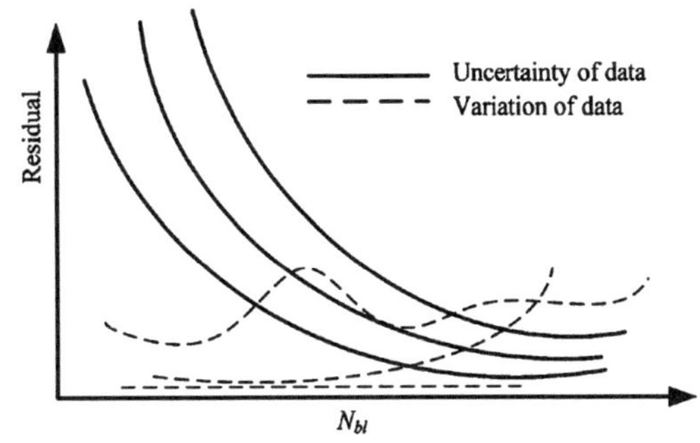

Fig. 3.11 Determination of the optimal buffer length [22]

buffer length. As a result, a method based on hypothesis testing is proposed to solve this problem.

It is meaningful to utilize all measurements to extract as much information as possible. That is, it is logical to process a buffer of PMU measurements instead of using only one measurement. It is difficult to describe the variation of PMU measurements in a probabilistic sense, and it is difficult to quantify the impact of the variation of data on state estimation. By applying a method based on hypothesis testing [23], the basic idea is to test whether the system is static or not.

The hypothesis testing technique is illustrated in ref. [22]. Since a buffer of PMU measurements is processed, the PMU error variance could be calculated. It is easy to notice that the two largest variances correspond to the two smallest buffer lengths. As conclusion, a larger variance results in a lower weight in the state estimation. According to this, from the developed procedure for the optimal buffer length determination the state estimation weights associated with PMU measurements could be determined.

It turned out that in reality buffer data shows that correlation exists in time between different buses. This measurement correlation represents useful information that can be included in the SE to obtain an optimal approach that improves accuracy. In ref. [24], a novel procedure is proposed to include the time and space correlation modeling in the PMUs and SE. These time series can be modeled by stationary vector autoregressive models in order to forecast or filter PMU measurements from different sensors.

The theoretical analysis and simulation results show the improved performance obtained with the proposed method over existing ones. The proposed approach improves the state estimation and provides more accurate confidence intervals by considering important information namely the space and time correlation available in the recorded PMUs resulting in a more robust state estimation.

Recently, further analysis of this interesting approach was performed [25]. It can be observed that the use of buffered phasor measurements provides performance enhancement in hybrid SE. The benefits of using variable buffer lengths over fixed buffer length values are also demonstrated.

3.7 Synchronous Generator Parameter Identification

Knowledge of generator operational parameters is vital for reliability stability studies and postmortem analysis. It is very important to have accurate models of the dynamic elements in the system, including synchronous machines and other power plant controllers, for decision-making during system planning and operations. Parameters of generators may differ from those stated in a database due to aging processes, magnetic saturation, or changes of temperature during machine operations [26].

Disconnecting of power plant elements for testing and model validation is neither technically practical nor economically attractive as it can reduce the robustness of the power system and may increase the overall generation costs. In addition, synchronous machine parameters may be obtained by off-line tests, but these tests cannot adequately address operating changes of the machine. Thus, online techniques become the best option for estimating and updating models of power system plants. PMUs can provide reliable information for validation purposes when located near or at the power plant.

For the estimation of the equivalent circuit parameters of large synchronous generators the equivalent circuit of Park is used. Park's model is effectively enabled through a transformation of *abc* quantities to an equivalent dq0 set of quantities that refers all quantities to a rotor frame of reference. Park's transformation is given by:

$$P = \sqrt{2/3}\begin{bmatrix} \dfrac{1}{\sqrt{2}} & \dfrac{1}{\sqrt{2}} & \dfrac{1}{\sqrt{2}} \\ \cos\theta & \cos\left(\theta - 2\pi/3\right) & \cos\left(\theta + 2\pi/3\right) \\ \sin\theta & \sin\left(\theta - 2\pi/3\right) & \sin\left(\theta + 2\pi/3\right) \end{bmatrix}$$

where the angle θ is given by:

$$\theta = \omega_{\mathrm{R}} t + \delta + \pi/2$$

ω_{R} is the rated (synchronous) angular frequency in rad/s and δ is the synchronous torque angle in electrical radians.

The mathematical model is derived in its expanded form with *abc* quantities:

$$\begin{bmatrix} v_a \\ v_b \\ v_c \\ -v_F \\ -v_D \\ -v_G \\ -v_Q \end{bmatrix} = -\begin{bmatrix} r_a & 0 & 0 & 0 & 0 & 0 & 0 \\ 0 & r_b & 0 & 0 & 0 & 0 & 0 \\ 0 & 0 & r_c & 0 & 0 & 0 & 0 \\ 0 & 0 & 0 & r_F & 0 & 0 & 0 \\ 0 & 0 & 0 & 0 & r_D & 0 & 0 \\ 0 & 0 & 0 & 0 & 0 & r_G & 0 \\ 0 & 0 & 0 & 0 & 0 & 0 & r_Q \end{bmatrix}\begin{bmatrix} i_a \\ i_b \\ i_c \\ i_F \\ i_D \\ i_G \\ i_Q \end{bmatrix} - \begin{bmatrix} \dot{\lambda}_a \\ \dot{\lambda}_b \\ \dot{\lambda}_c \\ \dot{\lambda}_F \\ \dot{\lambda}_D \\ \dot{\lambda}_G \\ \dot{\lambda}_Q \end{bmatrix} + \begin{bmatrix} v_n \\ 0 \end{bmatrix}$$

where r is the winding resistance, v is the voltage, i is the current, and λ is the flux linkage. Reformulating in dq0 components becomes:

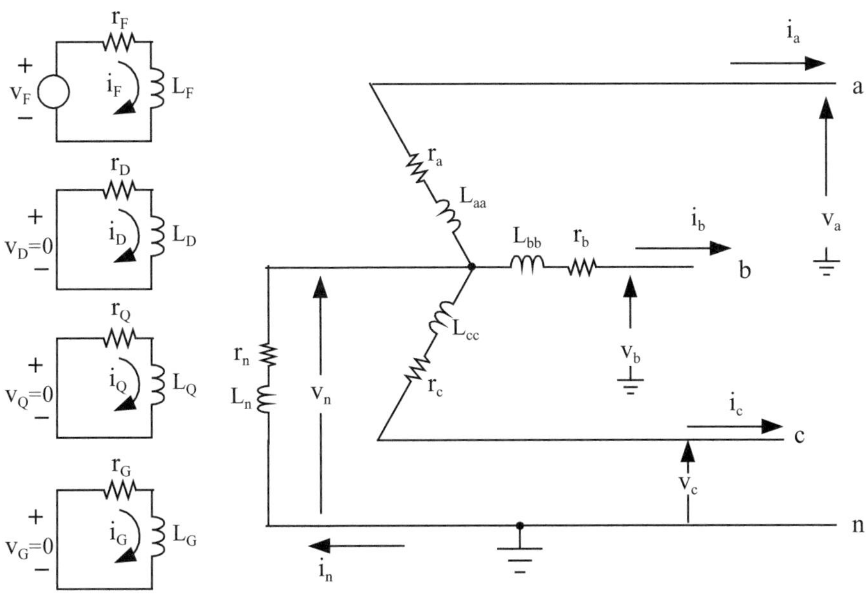

Fig. 3.12 Park's model of a synchronous machine [27]

$$
\begin{bmatrix} v_0 \\ v_d \\ v_q \\ -v_F \\ -v_D \\ -v_G \\ -v_Q \end{bmatrix} = -\begin{bmatrix} r+3r_n & 0 & 0 & 0 & 0 & 0 & 0 \\ 0 & r & \omega(L_{AQ}+\ell_q) & 0 & 0 & \omega L_{AQ} & \omega L_{AQ} \\ 0 & -\omega(L_{AD}+\ell_d) & r & -\omega L_{AD} & -\omega L_{AD} & 0 & 0 \\ 0 & 0 & 0 & r_F & 0 & 0 & 0 \\ 0 & 0 & 0 & 0 & r_D & 0 & 0 \\ 0 & 0 & 0 & 0 & 0 & r_G & 0 \\ 0 & 0 & 0 & 0 & 0 & 0 & r_Q \end{bmatrix} \begin{bmatrix} i_0 \\ i_d \\ i_q \\ i_F \\ i_D \\ i_G \\ i_Q \end{bmatrix}
$$

$$
-\frac{1}{\omega_B}\begin{bmatrix} L_0+3L_n & 0 & 0 & 0 & 0 & 0 & 0 \\ 0 & L_{AD}+\ell_d & 0 & L_{AD} & L_{AD} & 0 & 0 \\ 0 & 0 & L_{AQ}+\ell_q & 0 & 0 & L_{AQ} & L_{AQ} \\ 0 & L_{AD} & 0 & L_{AD}+\ell_F & L_{AD} & 0 & 0 \\ 0 & L_{AD} & 0 & L_{AD} & L_{AD}+\ell_D & 0 & 0 \\ 0 & 0 & L_{AQ} & 0 & 0 & L_{AQ}+\ell_G & L_{AQ} \\ 0 & 0 & L_{AQ} & 0 & 0 & L_{AQ} & L_{AQ}+\ell_Q \end{bmatrix} \begin{bmatrix} \dot{i}_0 \\ \dot{i}_d \\ \dot{i}_q \\ \dot{i}_F \\ \dot{i}_D \\ \dot{i}_G \\ \dot{i}_Q \end{bmatrix}
$$

where all quantities are in per unit except ω_B which is in rad/s and time which appears in the derivative terms in seconds. The v and i terms are taken as measured inputs and several elements of the coefficient matrices are to be estimated. Figure 3.12 shows the model used in this work, and which consists of three stator windings, one field winding and three damper windings.

The concept of extended Kalman filter [28] is applied in the process of parameter estimation. It is feasible to identify some or all synchronous generator constants using Kalman filter estimation provided that the sampling rate of process measurements (all three phase currents, all three phase terminal voltages, the DC field voltage and current) is sufficient.

Also, testing of estimated parameters using online machine data during system disturbances is performed. It is shown that the method can provide estimates of machine parameters with reasonable accuracy when a sufficiently large disturbance is present. A reasonable figure for accuracy of Kalman estimation of synchronous generator parameters is in the 1 % range (maximum error) for sampling rates at or faster than one sample per cycle. When PMU data are used, a faster sampling rate needs to be used for sufficient accuracy in the estimated parameters. The higher the sampling rate, the better the estimates. Thus, the suggestion is, instead of PMU data with industry standard 30 samples per second to go with 60 or 120 samples per second or even use a Digital Fault Recorder (DFR), capable of sub-cycle sampling [27].

3.8 Conclusions

The usage of synchrophasor technology for wide-area measurements, monitoring, analysis and control is enhancing electric power system reliability, efficiency, and resilience. PMUs enable finer and faster control of the network. A growing community of researchers and utility experts are working on advanced applications software to analyze and display synchrophasor data. Most of these applications focus on: diagnosing in real-time grid conditions and precursors that could lead to disturbances, and identifying and developing ways to mitigate or prevent evolving grid events and collapse; increasing grid asset utilization and system efficiency by expanding grid throughput and operating closer to the margin reliably and securely; enabling reliable and secure dynamic grid operation with changing resources mix, including integration of variable renewable resources, demand response, and variable load characteristics. All of this is a paradigm shift and moves the analysis from static analysis to dynamic analysis.

In the future, it is necessary to work with appropriate standard-setting organizations to complete the development and adoption of PMU devices and network standards and performance requirements to permit effective data exchange, and synchrophasor system and device interoperability and conformance.

Prof. Heydt's numerous research contributions span several topics and have changed the way we think about many topics in the power engineering field. In this chapter just a few contributions in the synchrophasor measurements arena are given. For a man with extraordinary technical and intellectual curiosity, such as Prof. Heydt, promising synchrophasor measurement applications in power engineering triggered extreme interest and involvement resulting in his significant contributions to this revolutionary technology.

References

1. Phadke AG (2002) Synchronized phasor measurements-a historical overview. In: Transmission and distribution conference and exhibition 2002: Asia Pacific, vol. 1. IEEE/PES, pp 476–479
2. Lewandowski W, Azoubib J, Klepczynski WJ (1999) GPS: primary tool for time transfer. Proc IEEE 87:163–172
3. IEEE Standard (1995) IEEE standard for synchrophasor for power systems – IEEE Std. IEEE, Piscataway, NJ, pp 1344–1995
4. IEEE Standard (2006) IEEE standard for synchrophasors for power systems – IEEE Std C37.118-2005. IEEE, Piscataway, NJ
5. IEEE Standard (2011) C37.118.1-2011 – IEEE standard for synchrophasor measurements for power systems. IEEE, Piscataway, NJ
6. IEEE Standard (2011) C371182-2011 – IEEE standard for synchrophasor data transfer for power systems. IEEE, Piscataway, NJ
7. North American Reliability Corporation. www.nerc.com.
8. Phadke AG, Thorp JS (2008) Synchronized phasor measurements and their applications. Power electronics and power system. Springer, New York, NY
9. Kosterev DN, Taylor CW, Mittelstadt WA (1999) Model validation for the August 10, 1996 WSCC system outage. IEEE Trans Power Syst 14(3):967–979
10. Denegri GB, Invernizzi M, Milano F (2002) A security oriented approach to PMU positioning for advanced monitoring of a transmission grid. In: International conference on power system technology (PowerCon 2002), vol. 2, pp 798–803
11. Zivanovic R, Cairns C (1996) Implementation of PMU technology in state estimation: an overview. In: 4th IEEE AFRICON, vol. 2, pp 1006–1011
12. Ingram SBM, Matthews S, Meliopoulos AP, Cokkinides G (2004) "Use of phasor measurements, SCADA and IED data to improve the state estimation. 7th Fault and Disturbance Analysis Conference, Atlanta, GA
13. Rice M, Heydt GT (2006) Power system state estimation accuracy enhancement through the use of PMU measurements. In: Proceedings of the IEEE PES T&D conference 2006, Dallas, TX, pp. 1460–1466
14. Rice M, Heydt GT, Jiang W, Vittal V (2008) Design of state estimator measurements based on condition indicators, Electr Pow Compo Syst 36(7):665–679
15. Jiang W, Vittal V, Heydt GT (2007) A distributed state estimator utilizing synchronized phasor measurements. IEEE Trans Power Syst 22(2):563–571
16. Haughton D, Heydt GT (2013) A linear state estimation formulation for smart distribution systems. IEEE Trans Power Syst 28(2):1187–1195
17. Zhang Q, Vittal V, Heydt GT, Chakhchoukh Y, Logic N, Sturgill S (2012) The time skew problem in PMU measurements. IEEE Power and Energy Society General Meeting, San Diego, CA
18. Brown RG, Hwang PYC (1997) Introduction to random signals and applied Kalman filtering: with MATLAB exercises and solutions. Wiley, New York, NY
19. IEEE Standard (2008) C5713-2008 – IEEE standard requirements for instrument transformers. IEEE, Piscataway, NJ
20. American national standard requirements for power-line carrier coupling capacitors and coupling capacitor voltage transformers (CCVT), ANSI C93.1-1999, 1999
21. Zhang Q, Vittal V, Heydt GT, Logic N, Sturgill S (2011) The integrated calibration of synchronized phasor measurement data in power transmission systems. IEEE Trans Power Deliver 26(4):2573–2581
22. Zhang Q, Chakhchoukh Y, Vittal V, Heydt GT, Logic N, Sturgill S (2013) Impact of PMU measurement buffer length on state estimation and its optimization. IEEE Trans Power Syst 28 (2):1657–1665

23. Shanmugan KS, Breipohl AM (1988) Random signals: detection, estimation and data analysis. Wiley, New York, NY
24. Chakhchoukh Y, Vittal V, Heydt GT (2014) PMU based state estimation by integrating correlation. IEEE Trans Power Syst 29(2):617–626
25. Murugesan V (2013) Error detection and error correction for PMU data as applied to state estimators. Master thesis, Arizona State University – School of Electrical, Computer and Energy Engineering
26. Kyriakides E, Heydt GT, Vittal V (2004) On-line estimation of synchronous generator parameters using a damper current observer and a graphic user interface. IEEE Trans Energ Convers 19(3):499–507
27. Lin K, Kyriakides E, Heydt GT, Logic N, Singh B (2010) Experience with synchronous generator parameter identification using a Kalman filter. IEEE Power and Energy Society General Meeting, Minneapolis, MN
28. Tumageanian A, Keyhani A (1995) Identification of synchronous machine linear parameters from standstill step voltage input data. IEEE Trans Energ Convers 10(2):232–240

Chapter 4
Renewable Resource Reliability and Availability

Gerald B. Sheblé

4.1 Introduction

Wind generators are complex systems based on the latest aerodynamic, mechanical, and electrical designs incorporating coordinated sophisticated control systems. Wind generators have been erected in increasing numbers in the USA, the European Union, China, and other locations, especially smaller islands with no interconnections. European companies and US companies have lead developing technology.

Solar cells are the least complex systems based on the latest manufacturing processes with sophisticated conversion systems to turn direct current into alternating current at the highest efficiency.

This work is concerned with the basic reliability and availability analysis of wind and solar generation. All renewable resources are demonstrated in this work. The role of probabilistic production costing (PPC) to determine the reliability and the expected costs is several decades in evolution [1–13]. Indeed, most tariff cases require the evaluation of the reliability of the power system by PPC to find the loss of demand probability (LODP) and the expected energy not served (EENS).

Renewable energy supply (RES) generation penetration in electricity systems is expected to demonstrate an explosive growth in the near term. Power systems operation procedures, tariffs, and legal contracts will be modified to integrate the specific characteristics of variable renewable energy resources with unique operation characteristics (wind turbines, hydro, pumped hydro, photovoltaic, etc.). A number of interconnection issues arise requiring development of new generation expansion planning methodologies. Such resources are distributed in nature, yet require transmission capabilities to provide the same level of reliability when a

G.B. Sheblé
Portland, OR, USA
e-mail: gbsepmt@gmail.com

E. Kyriakides et al. (eds.), *Electric Power Engineering Research and Education*,
Power Electronics and Power Systems, DOI 10.1007/978-3-319-17190-6_4

large scale penetration of renewable energy is eminent. A probabilistic approach is necessary due to the uncertainty of variable resources incorporating the statistics of the customer demand, wind, solar, and water availability. The non-dispatchable RESs are inherently not amenable to integration with generation. The penetration level of variable renewable energy resources has to be limited according to restrictions implied by the energy curtailment which can occur when the customer demand is low and RES generation is high. Hydro pumped storage plants, and other storage technologies, decrease this curtailment and consequently increase the penetration level of REGs. Fast reserve capacity is required to deal with the potentially large variations of variable REGs generation. The necessary storage and reserve capacity have to be coordinated and incorporated in the costs of different scenarios related to expansion planning or operation simulation. Grid costs related to the Transmission System Expansion have to be calculated and justified to maintain the reliability of service once the penetration of high REGs occurs.

This work reviews the state-of-the-art probabilistic production costing for renewable energy resources, especially solar cell and wind generation farms. A farm is a collection of devices aggregated into one virtual power plant. Each farm provides all of the resources as contracted to be as similar as possible to a conventional utility power plant. The tools presently in use are demonstrated, and recent developments of potential tools are outlined.

This chapter starts with consideration for each generation resource. The viewpoint is of an independent power producer or a horizontally integrated electric utility power plant. Both cases require the evaluation of all services provided by a generation resource. The first objective of this work is to present models based on existing data to predict the availability of renewable generation as a basis for power system reliability analysis and probabilistic production costing. The second objective is to present the similarities of wind generation with solar cells and with hydro generation. The third objective is to present the integration of renewable generation with the total generation portfolio to provide all energy and power services as required for modern energy system operation as expected by regulators and customers.

4.1.1 Independent Power Producer Interconnection Considerations

The interconnection requirements for generation have evolved since independent power producers (IPPs) were first enabled to connect to the grid. Wind and solar generation units are aggregated into "farms" as defined by a common transmission or distribution interconnection bus, thus a virtual power plant. It is commonly required/recommended that all new wind farms have capabilities similar to thermal generator functions [14]. Each is presented in the following as applies to renewable and to conventional power generation by coal, oil, gas, nuclear, or hydro.

Power systems require ancillary services to efficiently and reliably transport energy from supplier to buyer. Most of the ancillary services are traditionally supplied by generation. Thus, voltage regulation at point of connection similar to conventional power plants is required. Low voltage ride through (LVRT) without the power electronics withdrawing from service enhance reliability. Specified level of monitoring, metering, and event recording as a conventional power plant is often required. Power curtailment capability as a temporary limit as a transient or dynamic response is needed for inertial response to maintain stability. Ramping capability is required for demand following capability for both increases and decreases. Governor response capability is needed for frequency response to maintain power system stability. Reserve response capability is needed for unplanned resource, transmission, or demand changes. Voltage regulation is needed for transient, dynamic, and steady state power system control. Voltage regulation is needed for loss minimization.

4.1.2 Traditional Reliability Analysis Probabilistic Production Costing

The traditional reliability analysis used production cost models to calculate a generation system's production costs, cost-effective energy imports, profitable energy for sales to other systems, and fuel consumption. They are widely used throughout the electric utility industry as the cornerstone of long-range system planning, in fuel budgeting, and in system operation. The future system energy costs are computed by using computer models of expected demand patterns and simulating the operation of the generation and transmission system to meet these demands. Since generating units are not perfectly reliable and future demand levels cannot be forecast with certainty, production cost programs are based on probabilistic models and are used to compute the statistically expected need for energy storage, emergency energy and capacity supplies, or demand reductions.

The basic digital simulation of the generation system involves representation of:

- Generating unit efficiency characteristics including but not limited to input–output curves, ramp rate capabilities, governor response capability, start-up and banking cost curves, voltage response capabilities, and fuel costs per unit of energy supplied.
- System operating policies.
- Unit operation scheduling.
- Online unit economic dispatch constraints.
- Demand characteristics and capabilities.
- Contracts for the purchases and sales of both energy and power capability.

The first power system production cost models were deterministic, in that the status of all units and energy resources was assumed to be known and the demand was a single estimate [15].

Production cost programs involve modeling all of the generation characteristics and many of the controls discussed previously, including fuel costs and supply, economic dispatch, unit commitment, and hydrothermal coordination. They also involve modeling the effects of interchange and market transactions.

Deterministic programs incorporated operational generation scheduling techniques in a simulation model. In the most detailed of these, the online unit commitment program coupled with and optimum power flow might be used in an off-line study mode [16]. These are used in studying issues that are related to system operations such as purchase and sale decisions, transmission access issues and near-term decisions regarding operator-controlled demand management.

Stochastic production cost models are usually used for longer-range studies that do not involve near-term operational considerations. In these problem areas, the risk of sudden, random, generating unit failures and random deviations of the demand from the mean forecast are considered as probability distributions. Such models are beyond the scope of this work.

This work describes the basic ideas used in the probabilistic production cost models. This work reviews several, but not all, of the methods used for Probabilistic Production Costing (PPC). The timeline of the work is included generally in the following. However, the basic introductory texts include [17–23].

The pseudo-code block diagram shows the organization of a prototype energy production cost program. The computation simulates the system operation on a periodic basis with system data input being altered at the start of each interval.

Such programs take into account the need for scheduled maintenance outages. Algorithms are incorporated to simulate the maintenance outage allocation procedure actually used, as well as to process maintenance schedules that are input to the program.

Expansion planning and fuel budgeting production cost programs require demand models that cover weeks, months, and/or years. The expected demand patterns may be modeled by the use of typical, normalized hourly demand curves for the various types of days expected in each subinterval (i.e., month or week) or else by the use of demand duration or demand distribution curves. Demand models used in studying operational issues involving the next few hours, days, or weeks, are usually chronological demand cycles (Table 4.1).

The scheduling of unit maintenance outages [24] may involve time intervals as short as a day or as long as a year. The requirements for economic data such as unit, plant, and system consumption and fuel costs are usually on a monthly basis. When these time interval requirements conflict, the demand model must be created for the smallest subinterval involved.

Production cost programs are the risk engine for interchange contract selection process, energy storage scheduling (pumped hydro, compressed air, liquid compressed air, hydrogen, etc.), renewable resource scheduling (hydro, wind, solar cell generation, etc.). Previous energy management system (EMS) production cost

Table 4.1 Procedural diagram of single area energy production cost planning program

Initialize variables
Get data
Build demand duration curves for each period
Iterate to dispatch hydro, take or pay contracts, energy storage across periods
Modify expected demand requirements for maintenance, interchange, or multi-period energy storage
Schedule thermal resources and energy storage within period
Save results for reports

models are usually intended to produce short-term production costs computations (i.e., a day to 2 weeks) to facilitate negotiations for energy (or power) interchange, to generate market bids, risk management, or to compute cost savings in order to allocate economic benefits among pooled companies. Traditionally, such computations were deterministic using Economic Dispatch and Unit Commitment as outlined in [22]. The production cost simulation is used to evaluate costs under several scenarios. Interchange negotiations are the classic case of the make or buy decision. The system operators would evaluate the cost of producing the energy on the system versus the costs of purchasing it, sometimes on a daily basis.

Production cost computations are also needed for fuel budgeting. This involves making computations to forecast the needs for future fuel supplies at specific plant sites. Arrangements for fuel supplies vary greatly. The utility may control the mining of coal or the harvesting and transportation of natural gas; for others it may contract for fuel to be delivered to the plant. In many cases, the utility will have made a take or pay arrangement with a fuel supplier for the fuel needed for a specific plant. An alternative case, the generation company may have to obtain fuel supplies on the open (i.e., "spot") market at whatever prices are prevailing at that time. Thus, bids for the fuel markets, and hedging on other fuel markets, require analytic support for the positions taken. The take or pay fuel contract is one of the more interesting applications when the duration is over several weeks or months.

Operating center production cost needs typically had a 7-day time horizon to include the weekend demand trough. The fuel budgeting time span may encompass 1–5 years and might, in the case of the nuclear or mine-mouth coal plant studies, extend out for a decade. System planning expansion studies usually encompass a minimum of 10 years and in many cases extend to 30 years into the future. It is this difference in time horizon that makes different models and approaches suitable for different problems.

4.1.3 Basic Probabilistic Production Costing

Until the 1970s, production cost estimates were usually computed on the basis that the total generating capacity is always available, except for scheduled maintenance outages [15]. Operating experience indicated that the forced outage rate of thermal-generating units tended to increase with the unit size. Power system energy production costs are adversely affected by this trend. The frequent long-duration outages of the more efficient base-demand units required running less efficient, more expensive plants at higher capacities and for more hours and the import of emergency energy. Some utility systems reported the operation of peaking units for more than 150 h each month, when these same units were originally justified under the assumption that they would be run only over a few hours per month, if at all.

The use of convolution (Balériaux/Jamoule/De Duertechin/Booth) was a major breakthrough to include the availability and the forced outage rate two-state model of a unit to demonstrate why peaking units were required more than expected [4, 10, 17, 19, 25–31]. The convolution method was the first generally accepted technique to include forced outage rates used by all regulatory authorities. Booth [4, 25] demonstrated how to model thermal units, hydro units, and interchange with a major benchmark improvement of peaking unit use. There were several key papers which summarized, justified, and extended the convolution methods [7, 26, 27, 29, 30, 32–37]. There are several test cases (benchmarks) for testing each new method [13, 38, 39].

Production cost programs are mathematical methods based on probabilistic models of both the demand to be served and the energy and capacity resources. The models of the generation need to represent the unavailability of basic energy resources (i.e., hydro-availability), the random forced outages of units, and the effects of contracts for energy sales and/or purchases. The computation may also include the expected cost of emergency energy, which is sometimes referred to as the *cost of unsupplied energy or of emergency energy*, especially as a financial indicator to serve the EENS.

The basic difficulties that were noted when using deterministic approaches to the calculation of systems production cost were:

- The base-demand units of a system are demanded in the models for nearly 100 % of an interval.
- The midrange, or "cycling," units are demanded for periods that depend on their priority rank and the shape of the demand-duration curve.
- For any system with reasonably adequate reserve level, the peaking units were not committed nor dispatched above minimum generation.

These computations were known to be incorrect whenever random-unit forced outages occurred. The unavailability of thermal-generating units due to unexpected, randomly occurring outages is fairly high for large-sized units, with common values above 10 %. If the full forced outage rate is *q*, per unit, the particular generating unit is completely unavailable for 100*q% of the time it is supposed to be available. Generating units also suffer partial outages where the units must be derated (i.e., run at less than full capacity) for some period of time, due to the forced outage of some system component (e.g., a boiler feed pump or a fan motor). These partial forced outages may reach data reflecting a 25 % forced reduction in maximum generating unit capability for 20 % of the time.

Data on unit outage rates has been collected and processed in the USA by the National Electric Reliability Council (NERC) for several decades. The collection and processing of these data are important and difficult tasks.

There were two techniques used to handle the convolution of the demand distributions with generation capacity. The first was probability density functions of the demand using numerical convolutions where discrete values modeled all of the unit's distributions. The second included analytical methods that use continuous functional representations or the surrogate components which could be combined to reconstruct the original demand distribution. Such techniques ranged from statistical moments, cumulants, to Fourier components. Some analytical methods use orthogonal functions to represent both the demand and capacity probability densities [3, 4, 6, 30, 32, 40].

Early implementations treated fuel shortages, nonutility generation, and hydro as fixed energy resources. These were subtracted from the demand curve before processing the availability of the utility generation.

The first approach is the unserved demand distribution method (UDDM). The individual unit probability capacity densities are convolved with the demand distribution in a sequence determined by a fixed economic demanding criterion to develop a series of *unserved demand distributions*. Unit energy production is the difference between the unserved demand energy before the unit is scheduled (i.e., convolved with the previous unserved demand distribution) and after it has been scheduled. The demand forecast is the initial unserved demand distribution [22]. This work recognizes that this processing is in the backward direction in the dynamic programming sense.

The expected cost method (ECM) convolves the unit probability capacity densities convolved with each other in sequence to develop distributions of available capacity and the expected cost curve as a function of the total power generated. This work recognizes that this processing is in the forward direction in the dynamic programming sense. This expected cost curve may then be used with the demand distribution to produce the expected value of the production cost to serve the given demand forecast distribution.

The probabilistic production cost procedure uses thermal-unit heat rate models (i.e., heat input rate versus electric power output) composed of linear segments. This type of heat rate characteristic enabled the development of efficient probabilistic computational algorithm. The linear segments yield a stepped incremental cost curves. This model enabled an economic scheduling algorithm as any segment is fully demanded before the next is required. These unit input–output characteristics may have any number of segments so that a unit may be represented with many or only a few segments as appropriate for the studies at hand and the computational time requirements allowed. Merit Order Loading Economic Dispatch was one common approach [10, 41]. Unit thermal data are converted to cost per hour using fuel costs and other operating costs, as is the case with any economic dispatching technique. Unit commitment was usually approximated using an expected priority order based on economic or statistical analysis. Merit Ordering Commitment was often based on the historical order of commitment based on operational data. Once the unit commitment order was established, the various available demanding segments were placed in sequence, in order of increasing incremental costs. Finally, emergency resources (i.e., tie lines or pseudo tie lines) were placed last on the demand order list.

The essential difference between the results of the probabilistic procedure and the usual economic dispatch computations is that more units will be required if generator forced outages are considered.

Present implementations treat demand either as a cumulative probability distribution, where $P_n(x)$ is the probability of needing x MW, or as an hourly time series of demand. The following probabilistic procedure of thermal unit scheduling is the most easy to follow as it is based on decision analysis. "If there is a segment of capacity with a total of C (real power) available for scheduling, then denote q as the probability that x units of real power are unavailable (i.e., its unavailability) and p as the probability that x units of real power are available from this segment. After this segment has been committed, the probability of needing x units of real power or more is now $P'_n(x)$. Since the occurrence of demands and unexpected unit outages are statistically independent events, the new probability distribution is a combination of mutually exclusive events with the same measure of need for additional capacity:

$$P_n^{'}(x) = qP_{n-1}(x) + pP_{n-1}(x + C) \qquad (4.1)$$

where $q\ P_{n-1}(x)$ is the probability that new capacity C is unavailable times the probability of needing x, or more, generation, and $p\ P_{n-1}(x+C)$ is the probability C is available times the probability $(x+C)$. These two terms represent two mutually exclusive events, each representing combined events where x MW, or more, remain to be served by the generation system" [22].

This is a recursive computational algorithm used in sequence to convolve each unit segment with the resulting distribution of demand not served. As the next segment is dispatched, after the first segment is committed, the previous segment has to be deconvolved or removed from the EDDC.

It is obvious at this point that the process of enumerating each possible state in order to compute expected operation, energy generation, and unserved demands, is a "curse of dimensionality." These calculations cannot be implemented without an organized and efficient scheduling and dispatch method. For n units and m segments per unit, each of which may be on forced outage or available, there are m^n possible events to enumerate. Industrial practice has used up to fourteen levels of availability. The use of many levels is justified by historical use.

The data requirements for this approach include generation models, storage models, demand models, and procedures for unit commitment linked with either economic dispatch or optimal power flow.

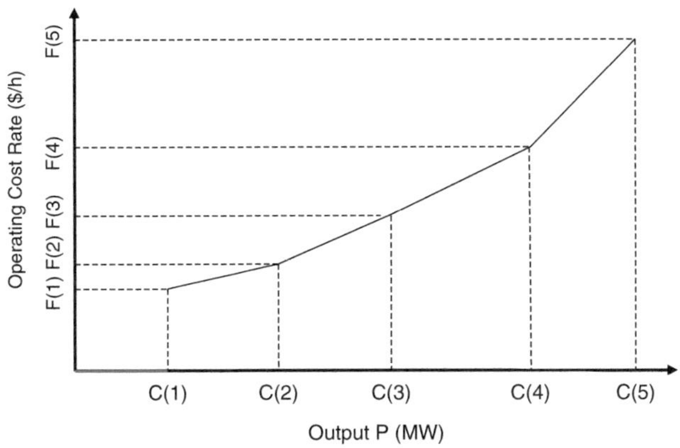

Fig. 4.1 Expected unit segments based on historical dispatch [23]

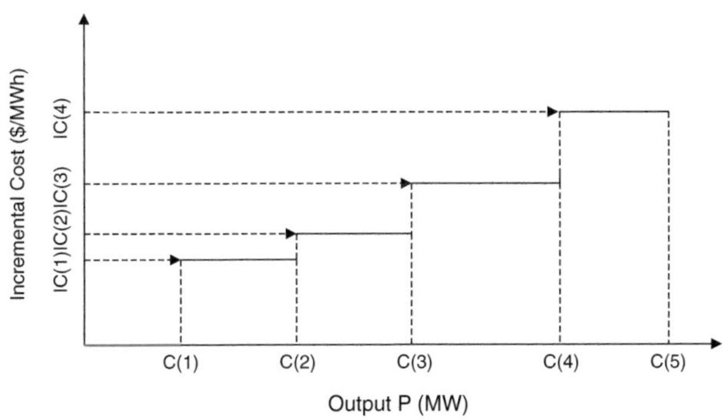

Fig. 4.2 Expected unit incremental segments based on historical dispatch [23]

4.1.4 Fossil Fuel Generation Reliability Models

The models for thermal generation modes may be as simple as a two state model to represent the availability or the forced outage of the unit. A more realistic model would include the expected dispatch levels for each unit based on historical use [23]. Consider the production segments shown in the following cost curves historically used for over three decades (Figs. 4.1 and 4.2).

A more realistic model would include the probability of each segment being dispatched. The following shows the unit distribution based on the previous operational history. Note that the sums of the probabilities must add to unity (1.0). Note for the two state model that the availability is easily related the failure probability.

$$p = (1 - q) \tag{4.2}$$

The reliability models for hydro generation are developed in a similar manner.

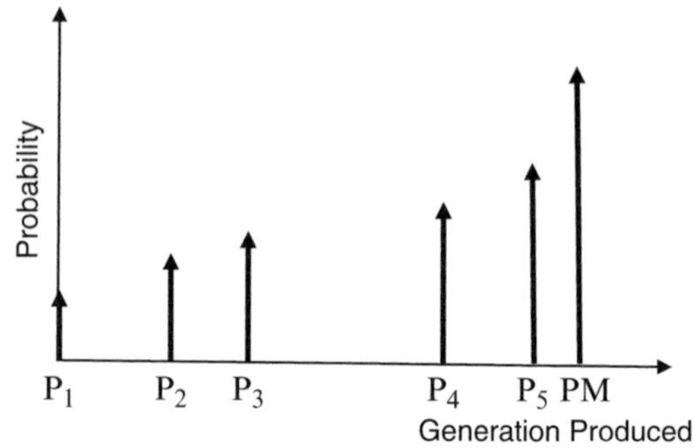

Fig. 4.3 Generation reliability model

However, hydro resources are not as controllable as fossil fuels. The availability of water is similar to the availability of wind and sun. The main difference is the energy collection area (watershed) is far larger than the blade length of a wind generator or the surface area of a solar cell. Thus, PPC needs to incorporate key concepts when fuel resources alter the reliability model due to availability issues (Fig. 4.3).

4.1.5 Fuel Availability

The treatment of renewable resources has been researched since the advent of probabilistic production costing [22, 24, 42]. The difference between renewable generation and traditional generation is the availability of fuel. Fossil fired generation enables control over the availability of fuel since it can be ordered in advance, stored in on-site inventory, or transported just in time. Renewable resources such as wind and sunlight cannot be ordered but can only be taken advantage when they are available. Thus, this work uses the term availability when dealing with raw resource presence for water, wind, and sunlight.

4.1.6 Renewable Energy Generation

The availability of water for hydro generation is also in question unless the reservoirs upstream are sufficiently large to store water for delayed release over all expected weather conditions. Run of river hydro is of the same uncertainty as wind or solar. Forecasting water inflows is dependent on the same variables, wind and sunlight, to thaw snow or to deposit rain in the drainage area upstream.

The availability of pumped hydro is certain as the water is moved from a lower reservoir to a higher reservoir as electric energy is available. It is only a matter of accounting the energy that could be stored at a lower cost period for regeneration at a higher cost period and the water lost due to evaporation or leakage [13].

The availability of wind generation requires an accurate forecast of the wind speed for every period of the study. Variations in wind speed can be accommodated by forecasting the distribution expected during the study period.

The availability of solar generation is similar to wind generation as an accurate forecast is required. The amount of sunlight, adjusted by cloud dispersion, is forecast for the study period, again as a distribution.

Water resources are scheduled normally as quickly as a day in advance as melt forecasts predict or as long as years when the storage is capable of maintaining the inventory over longer periods. Hydro resources are often constrained by irrigation, navigation, and flow restrictions. Such constraints represent firm long-term contracts in many instances. However, the importance of resource distributed storage, scheduling, and water shortages are adding to the constraints needed for actual production simulations. Hydro thermal coordination is beyond the scope of this work. This author has used PPC for Hydrothermal coordination for two Canadian utilities. The method was analogous to the maintenance scheduling method traditionally used [22].

4.1.6.1 Wind Generation Reliability Model

The reliability and economic impacts of wind generators are not represented as a two-state model for reliability evaluation. Renewable generation does not maintain a specified constant level. More complicated multi-state models are necessary [6, 8, 14, 38, 43–50] to include the availability of the "fuel" used. Wind forecasts are the critical input for estimating the amount of energy available.

The resulting power generated by a wind turbine can be expressed as:

$$p = \frac{1}{2} C_{\mathrm{P}} \rho V^3 A \tag{4.3}$$

where P power extracted from wind [W], C_{p} power coefficient, ρ air density (approximately 1.225 kg/m^3), V wind velocity [m/s], A swept area of rotor disc [m^2].

The typical relationship between the wind turbine power output and wind speed follow (Fig. 4.4).

The wind turbine state model is necessary for reliability evaluation in addition to the wind forecast. The wind turbine uncertainty is due mainly from uncertainties associated with the supply of the wind speed rather than the forced outage rate of the generator [47, 48, 51]. A common wind speed model is represented as a Weibull probability density function (pdf) as noted in these references.

The outage capacity pdf of the wind turbine can be obtained by combining the turbine's power output model and the wind speed pdf model as shown in the following. This yields a multi-state model because the wind speed is distributed widely. The probabilities of identical power are cumulated. Therefore, the state

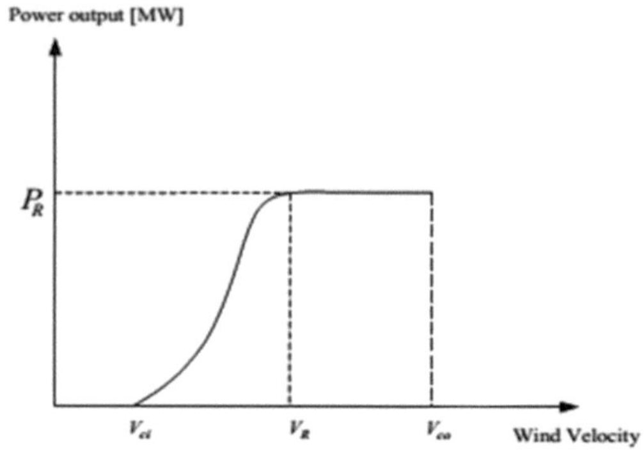

Fig. 4.4 Power output model of typical wind turbine generator [33]

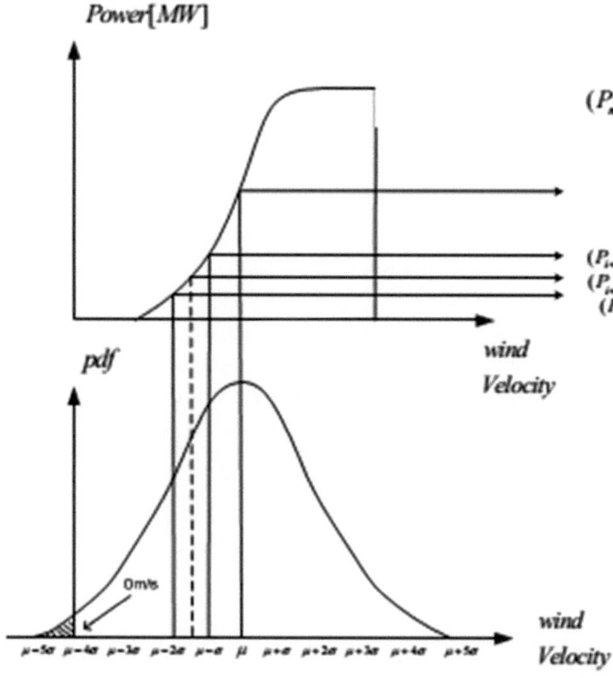

Fig. 4.5 Conversion of forecasted wind speed to extracted real power [33]

number is different from the wind speed band number of conventional generators (Fig. 4.5).

Using this multi-state model in reliability evaluation severely increase the convolution error propagation. As shown with traditional convolution methods, it is difficult and is computationally expensive due to step size-related reasons [52]. A simplification of the model is therefore needed.

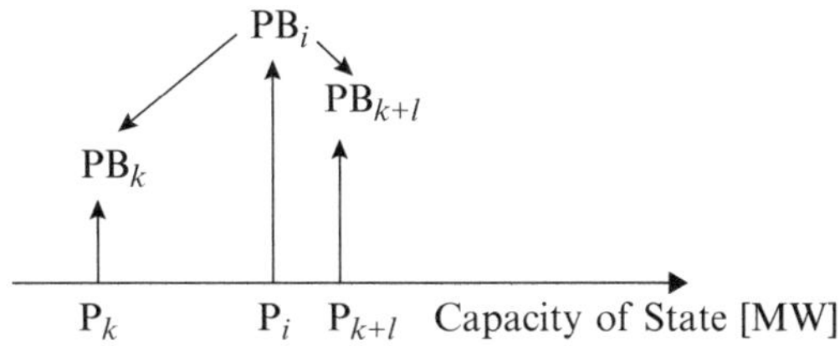

Fig. 4.6 Wind generator state model [33]

4.1.6.2 Reduced Availability Model

This work uses a probability tree to reduce the computational requirements for wind and solar generation. A simplification of the multi-state model is used with linear rounding sharing the ratio of probability linearly. The rounding method is presented graphically in the following figure and mathematically described by the following:

$$\begin{aligned} \mathrm{PB}_k &= \left(\frac{P_{k+1} - P_i}{\Delta P}\right)\mathrm{PB}_i \\ \mathrm{PB}_{k+1} &= \left(\frac{P_i - P_k}{\Delta P}\right)\mathrm{PB}_i \\ \Delta P &= P_{k+1} - P_k \end{aligned} \tag{4.4}$$

where P_i real power for the mean or average output [MW], P_k real power for the lower output [MW], P_{k+1} real power for the upper output [MW], PB_i probability for the mean or average power segment, PB_k probability for the lower power segment, PB_{k+1} probability for the upper power segment, k state number of the simplified multi-state model i, i state number of the original multi-state model.

The wind generation unit reliability, such as gear train failure, would have to be added to this model. This work uses the two state model of available or forced outage as a single unit as presented. When a wind farm is modeled in this fashion, then a more detailed model of mechanical forced outages is needed to represent the various mechanical failures to represent the wind "farm" as one unit. A wind "farm" could consist of any number of wind generators connected to the grid at a single transmission or distribution location (Fig. 4.6).

An alternative is to use a probability tree to round the outage capacity model. This approach is also used for incorporating errors in the demand forecast as presented below.

4.1.7 Solar Cell Model

The general mathematical expression for the solar cell power curve of a solar farm is shown in the following [4, 39, 53, 54] (Fig. 4.7).

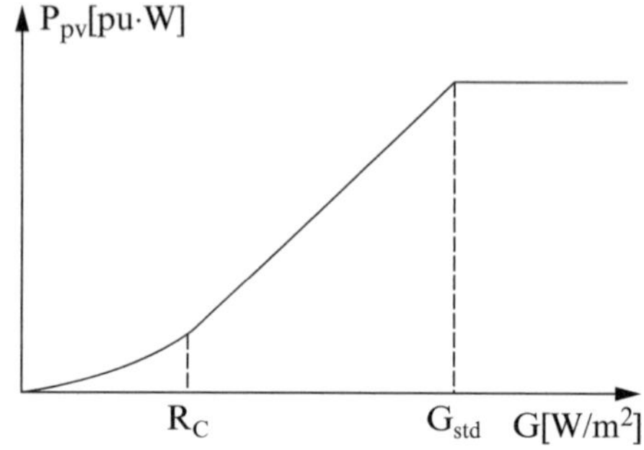

Fig. 4.7 SCG power model [33]

The power $P_{bi}(G_{bi})$ for a solar radiation G_{bi} band i ($i = 1,\ldots, N_b$) can be obtained using the following:

$$\begin{aligned} P_{b1}(G_{bi}) &= P_{sn}\left(G_{bi}^2/(G_{std}R_c)\right), 0 \leq G_{bi} \leq R_c \\ &= P_{sn}(G_{bi}/G_{std}), R_c \leq G_{bi} \leq G_{std} \\ &= P_{sn}, G_{bi} > G_{std} \end{aligned} \tag{4.5}$$

where N_b being the total number of bands, P_{bi} solar radiation-to-energy conversion function for band i of the SCG [MW], G_{bi} forecasted solar radiation at band i [W/m^2], G_{std} solar radiation in the standard environment set usually as 1,000 [W/m^2], R_c a certain radiation point set usually as 150 [W/m^2], P_{sn} equivalent rated capacity of the SCG [MW].

The solar farm operation state model needed for power system reliability evaluation is similar to the wind model as hardware forced outages are included.

The SCG outage capacity distribution is a combination of the SCG's power output model and the solar radiation model as shown in Fig. 4.8. This is a multistate model as solar radiations are distributed widely. The probabilities of identical power generated are cumulated as shown. Therefore, the state number is different from the solar radiation band number.

In this work, the reliability indices for systems including SCG are evaluated using the multi-state model for the SCGs and the two-state model for the conventional generators to account for forced outages of the cells.

4.1.8 Demand Models

The original demand model was the deterministic demand cumulative curve [15]. This model is obtained by ordering the demand into decreasing order. This model was used to find the ED for each period (hour) to find the production cost. The uncertainty of generation, demand, and constraints were not included. This

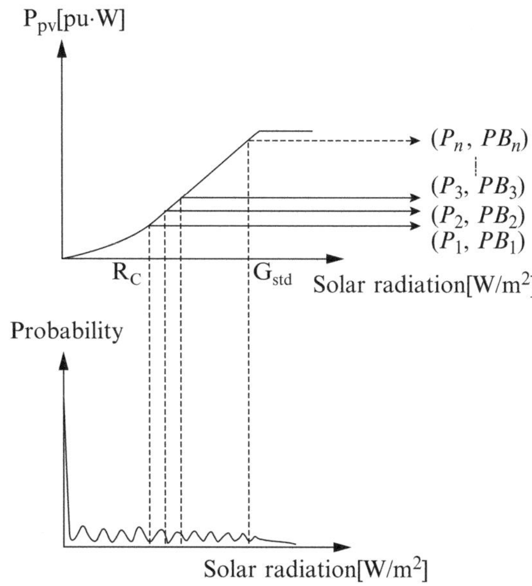

Fig. 4.8 Solar farm reliability model [33]

work calls this the Deterministic Demand Duration Curve (DDDC). This is obtained by aggregating the demand for each hour in order of increasing values. The resulting curve was used as the demand for ED or an equivalent UC. This curve is shown in the following (Figs. 4.9 and 4.10).

The number of hours for each segment is multiplied by the production costs and summed over all segments to find the production cost for the period under study. Typically, the period of study is a week to include the possibilities of energy storage over the weekend trough. This is the deterministic model for this work when the duration curve is sufficient demand differentiation.

The resulting curve yields a cumulative distribution when the time axis is converted to percentage of time. We use the Demand Cumulative Distribution Curve (DCDC) when we are using probabilistic scheduling and dispatch as in the following (Fig. 4.11).

The problem with both of these curves is that the time dependency is removed in this process. If the daily demand curves are needed for energy storage in the nightly troughs or due to the time of energy availability, the trough and the peak periods must be separated. It is this author's opinion that then only the hourly duration curve should be used.

The *demand cumulative duration curve* expresses the period of time (say number of hours) in a fixed interval (day, week, month, or year) that the demand is expected to equal or exceed a given generation (megawatt) value. It is usually plotted with the demand on the vertical axis and the time period on the horizontal axis as a histogram as shown in the above.

In representing future demands, sometimes it is satisfactory to specify only the total energy generation for a period. This is satisfactory if only total fuel

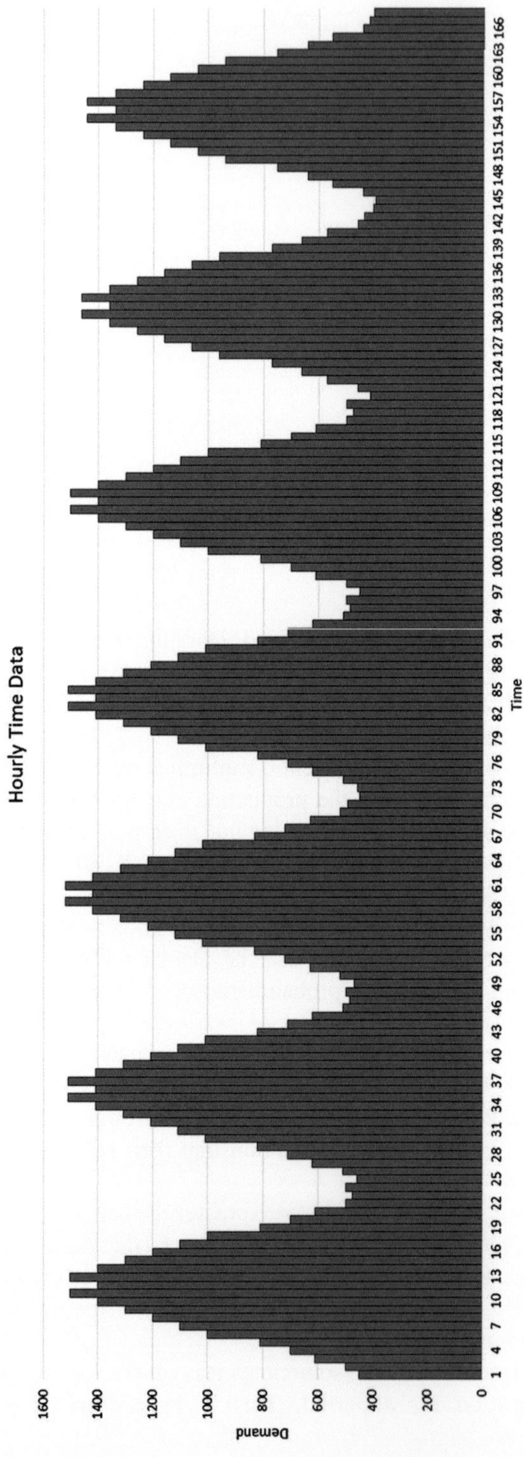

Fig. 4.9 Hourly demand curve (HDC) [70, 71]

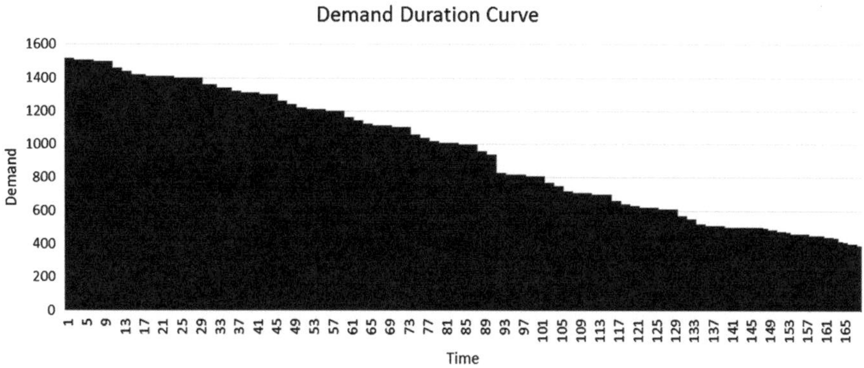

Fig. 4.10 Deterministic demand duration curve (DDDC)

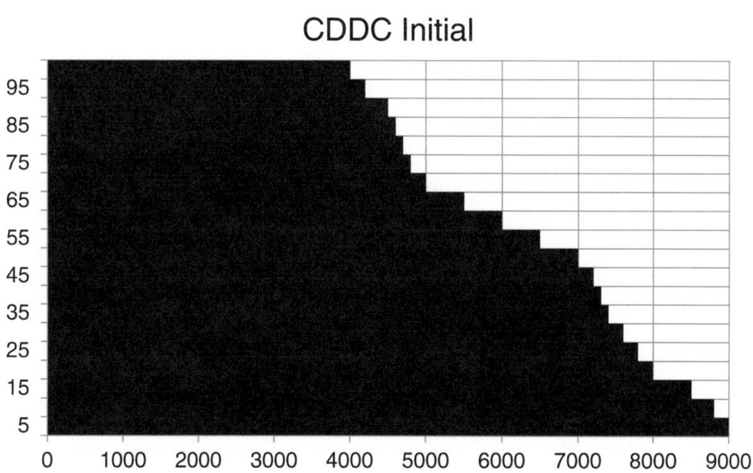

Fig. 4.11 Demand cumulative distribution curve (DCDC)

consumption and production costs are of interest and chronological effects are not important.

The demand is expressed as a discrete table in this work for both the probabilistic and the deterministic cases. The table needs to be only as long as the maximum demand divided by the uniform demand interval size used in constructing the table. It is computationally efficient to think in terms of regular discrete steps and recursive algorithms. Various demand-duration curves for the entire study period are arranged in sequence as is required to reach the study horizon for the commitment and dispatch algorithms. This classical definition of the convolution method is to show continuous curves for the demand model. Since the data is discrete upon acquisition, the discrete approach is used in this work.

4.1.9 Production Costing Demand Cumulative Duration Curve

A demand cumulative duration curve (DCDC) expresses the period of time (say number of hours) in a fixed interval (day, week, month, or year) that the demand is expected to equal or exceed a given generation (megawatt) value. It is usually plotted with the demand on the vertical axis and the time period on the horizontal axis as a histogram. The DCDC model was described previously.

This section's graphs show fewer demand bands for clarity of presentation. Since the unit availability model is convolved against this curve, we term this the equivalent demand duration curve (EDDC) to be consistent with other works.

In this table, $p(x)$ is the demand density function: the probability that the demand is exactly x MW and $P(x)$ is the demand distribution function; the probability that the demand is equal to, or exceeds, x MW.

The commitment and dispatch of the first segment is shown in the subsequent figure. The grey area shows the generation expected from the first unit segment. The dark area is the demand to be served by all remaining unit segments if the first unit segment is available.

The result of the convolution process is shown in the subsequent figure. This is the demand to be served by the remaining unit segments if the first unit segment is not available. Thus, the energy not supplied has to be supplied by the remaining units. The amount of energy generated by each unit is equal to the area under the demand-duration curve between the demand levels in megawatts supplied by each unit.

This process is repeated for each unit segment. Note that the first segment has to be deconvolved when the second segment is dispatched. The second segment has to be deconvolved when the third segment is dispatched. The process is the same for each subsequent segment.

In the simulation of the unit commitment/economic dispatch procedures with this type of demand model, thermal units are block-demanded. This means the units (or major segments of a unit) on the system are ordered in some fashion (usually average or expected cost) and are assumed to be fully demanded, or demanded up to the limitations of the demand-duration curve. The units are considered to be committed/dispatched in a sequence determined by their average cost. Sometimes the average cost at full demand [$/MWh] is used (Figs. 4.12, 4.13 and 4.14).

Besides representing the thermal generating plants, the various production cost programs must also simulate the effects of hydroelectric plants with and without water storage, take or pay fuel contracts, interchange or market contracts for energy and capacity purchases and sales, and pumped-storage hydroelectric plants. The original model of these resources was to modify the demand to be served by insertion into the EDDC when the energy matched the energy provided by that resource. This resulted in an approximation since the EDDC may not be able to use all of the energy given the block dispatch method. Alternatively, the availability of these resources should be included just as wind and solar availability is modeled.

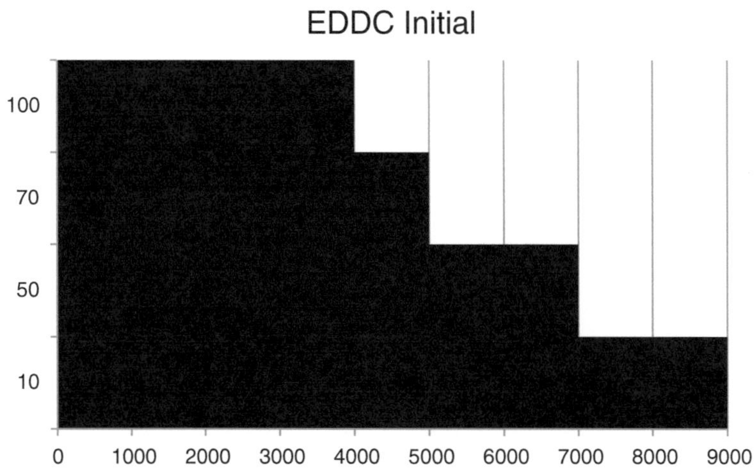

Fig. 4.12 EDDC

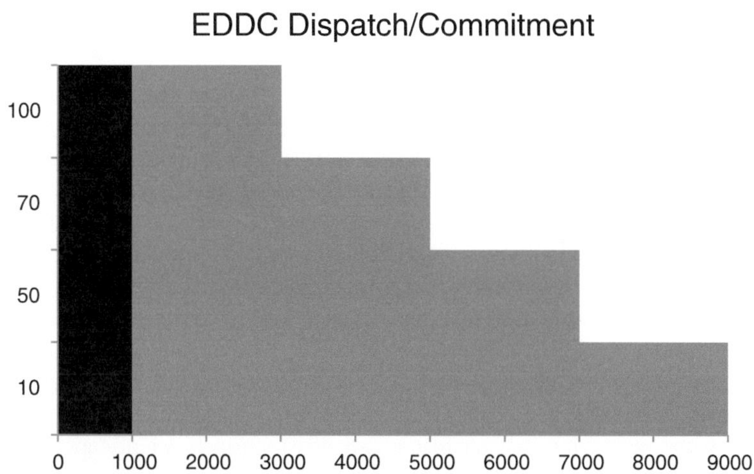

Fig. 4.13 EDDC unit 1 dispatch/commitment

The scheduling of the thermal plants should be simulated to consider the security practices and policies of the power system as well as to simulate, to some appropriate degree, the economic dispatch, commitment, and security constrained dispatch procedures used by the energy management system (EMS) to control the unit output levels.

More complex production cost programs used to cover shorter time periods may duplicate the logic and procedures used in the operation and control of the units. The most complex involve procedures such as economic dispatch, unit commitment, and hydrothermal scheduling. These programs will usually use hourly

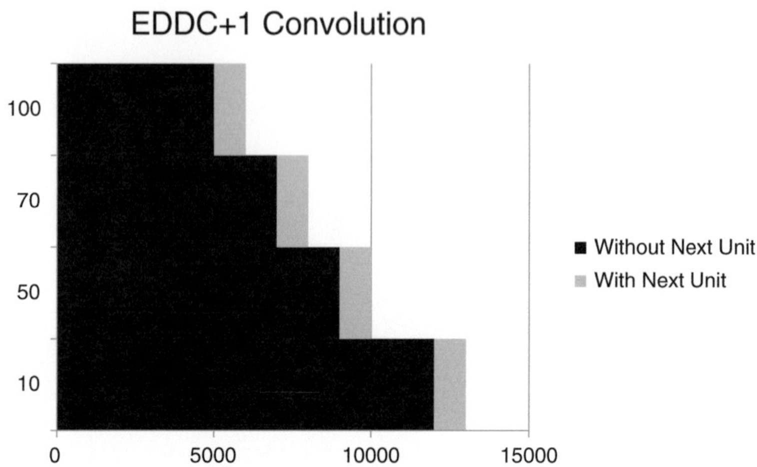

Fig. 4.14 EDDC($k+1$) convolution

forecasts of energy (i.e., the "hourly integrated demand" forecast) and thermal generating unit models that include incremental cost functions, start-up costs, and various other operating constraints.

The demand table above has been created for uniform demand-level steps of 20 MW each. The table also introduces the notation that is useful in regarding the demand-duration curve as a probability distribution. The demand density and distribution functions, $p(x)$ and $P_n(x)$, respectively, are probabilities. The distribution function, $P_n(x)$, and the density, $p(x)$, are related. The distribution function $D(x)$, also called the cumulative distribution function (CDF) or cumulative frequency function, describes the probability that a random variable X takes on a value less than or equal to a number x. The distribution function is sometimes also denoted $F(x)$.

As previously noted, it is interesting that many methods convert the discrete demand model into a continuous demand model through a curve fit procedure. Not only is this convenient for convolution description, it is advantageous for wave analysis techniques such as transforms and distribution moment representations [17, 22].

The distribution function is related to a continuous pdf, $P(x)$, by the following.

$$D(x) = P(X \leq x) \tag{4.6}$$

$$D(x) = \int_{-\infty}^{x} P(\xi) d\xi \tag{4.7}$$

So $P(x)$, when it exists, is simply the derivative of the distribution function:

$$P(x) = D^{'}(x) \tag{4.8}$$

Similarly, the distribution function is related to a discrete probability $P(x)$ by

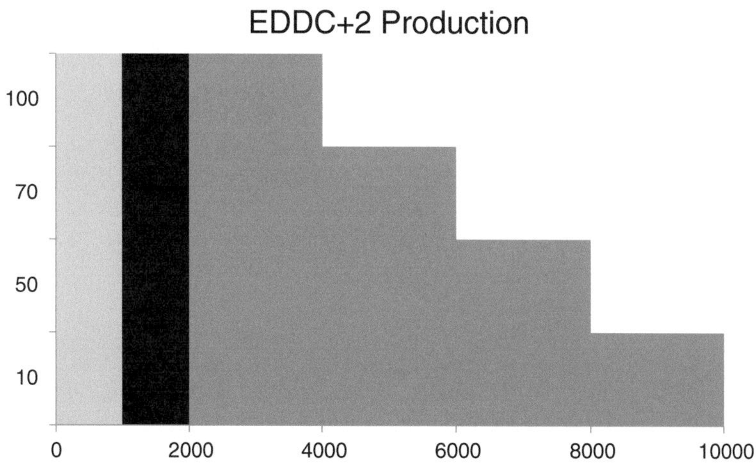

Fig. 4.15 EDDC($k+1$) unit 2 committed/dispatched

$$D(x) = P(X \leq x) \tag{4.9}$$

$$D(x) = \sum_{X \leq x} P(x) \tag{4.10}$$

For discrete-density functions (or histograms) in tabular form, it is easiest to construct the distribution by cumulating the probability densities from the highest to the lowest values of the argument (the demand levels).

The demand-duration curve is shown in a way that is convenient to use for the development of the probabilistic scheduling methods as follows (Fig. 4.15).

The cost rate for each unit is a linear function of the power output, *P*, for this work. *F*(*P*) is given by the cost at zero output plus the incremental cost rate times the power output (*P*).

Units are committed when the first block is dispatched. Unit 1 is used first because its average cost per MWh is lower. Unit 1 is online for 100 % and is generating at maximum generation.

4.1.10 Energy Demand Commodity Segmentation

Electric Energy is not as obviously defined as a commodity [11]. The reregulation to remove the implicit contracts in a vertical utility is extensive as detailed contracts must include the same reliability and quality of the vertically integrated utility. The energy is the main commodity traded. All of the remaining components are traded as ancillary contracts. The daily demand curve is segmented into hourly trading periods. Each period is broken into the commodity products to be traded through

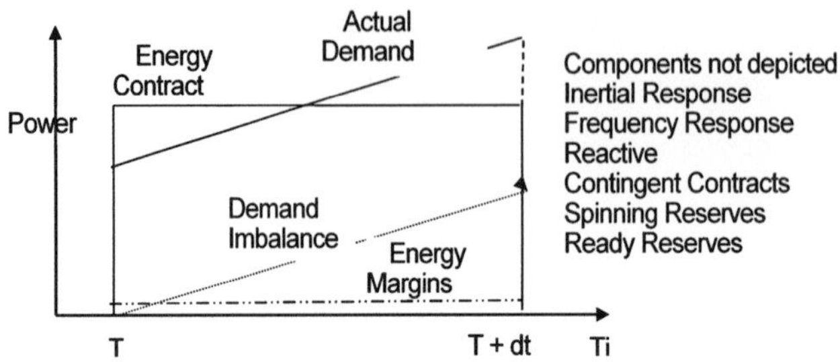

Fig. 4.16 Energy commodities [16]

markets. The first step is to identify the duration of the electric energy to be traded. This has traditionally been the hour in the USA. New Zealand had opted for the 5 min interval which has some advantages and disadvantages. The traditional hour (*dt*) is segmented as shown in the following (Fig. 4.16).

The actual demand curve is shown as a triangle for simplicity in the Energy commodities Figure. The energy rectangular block is the main energy contract traded to meet buyer demand. Since the energy block does not follow the actual demand, the hourly difference may be increasing or decreasing. This difference is traded on an Energy Imbalance Market (EIM) and may include more or less supply if the forecast error is significant. The above figure shows an increase in demand that was not expected during the energy block trading market. The Reactive Power demand is determined by a separate market. Reactive power is scheduled to minimize losses and/or to provide voltage support. The Frequency Response is a separate market as is the inertia market to exactly balance the supply and demand at each instant of time. The contingent market includes spinning reserve and outage contracts to supply the demand when one of the other markets (energy, imbalance) have a contract that is not deliverable due to supplier outage. Another cause for contingent contracts is forecast error if the EIB cannot react quickly.

If the market bids are based on marginal pricing, then an additional auction is needed for the fixed costs, such as capital budgeting. The use of incremental or expected average costs for bidding is beyond the scope of this work.

The various commodity inaccuracies are first due to forecast errors. One report [14] found the following error statistics. "The error in day ahead forecasts have standard deviation of 12 % or 400 MW on 3,300 MW generation. Error in hour ahead wind generation forecast were found to have standard deviations of approximately 4.2 % or 145 MW on 3,300 MW generation. Clearly one key component is wind forecasting per wind farm. The demand following (ED) 5 min intervals demonstrated a 3 % or 1.8 MW on 54.4 MW. AGC (frequency following) issues power commands on 6 s intervals It was found that NYISO needs 225–275 MW. The standard deviation of 6 s variability due to demand alone is 71 MW. Existing regulation practices suggest that 3 sigma (36 MW) would cover 99.7 % of the time. Spinning reserves required to cover the largest single contingency presently requirements is 1,200 MW on 3,300 MW generation."

This study [14] used GE's MARS program to find these requirements. MARS generated the LOLP and EENS (EUE). The MARS program simulated power system expansion using Monte Carlo techniques coupled with power system analysis programs to compute the operational margins needed in future periods of time based on forecasted demand. It was found that the unit effective capacities (UCAP) of inland wind generation were about 10 % of rated capacities, energy capacity factors are typically on the order of 30 %. The method to calculate effective capacities is covered in the following.

System stability was studied to find the disturbance and then the requirements to avoid instability. One of the key factors for wind and solar generation is the use of power electronics to convert DC to AC. Such conversion equipment requires a stable voltage signal to continue the process. One key characteristic is the need for sufficient voltage signal characteristics for the renewable energy source equipment to operate in unison with the power system of transmission and distribution components. LVRT critical (15 % voltage at point of interconnection) to improve system response to disturbances, ensuring faster voltage recovery and reduced post-fault voltage dips for 625 ms. This was consistent with FERC NOPR on wind generation interconnection requirements.

Energy balancing market or energy imbalance market (EIM) is of particular interest for the balancing of supply and demand. The AC transmission grid operates on an instantaneous balancing of supply and demand. It is the use of mechanical stored energy (inertia) which is the first response to changes in demand. FERC order 888 allows imbalance penalties to be applied to generators that operate outside of their schedule. There is no penalty for over generation. The under generation penalty is assessed at 150 % market rate spot price or $100/MWh. This is an anti-gaming rule for traditional generation. This has an impact on renewable generation whenever the wind or sun intensity unexpectedly reduces. Note that increases in renewable output are immediately stored in the system inertia until the frequency exceeds a stability target when generation and demand is disconnected. It is easier to turn down the supply than to increase it in a competitive marketplace.

4.1.11 Probabilistic Production Costing Methodologies

Reference [7] presents a concise comparison of six different probabilistic production cost simulation methods based on demand duration curves. The six methods compared included:

1. Piecewise Linear Approximation (PLA) method.
2. Segmentation method.
3. Equivalent Energy Function (EEF) method.
4. Cumulants method.
5. Mixture of Normal Approximation (MONA) method.
6. Fast Fourier Transform (FFT) method.

The PLA method used straight line segments to approximate the continuous EDDC. It is more closely related to the change in slope between the horizontal bars of the histogram called the EDDC. This approximation did lead to shorter computation times at the expense of reducing the granularity of the EDDC. The segmentation method divided the EDDC into a more coarse representation of the EDDC. Again this did reduce the computational effort of the convolution technique. The EEF and the Cumulants methods used another approximation to the EDDC by reducing the curve to a fixed set of components. The cumulants method is based on the moments of the equivalent demand. The MONA is another approximation based on mixtures of normal to represent the EDDC and the supply system to take advantage of these distribution properties. The FFT algorithm is the approximation of the EDDC by a Fourier series. The attraction of all such techniques is that the convolution or deconvolution involves only the addition or subtraction of the components. However, any inaccuracy of representation of the EDDC carries through to the results of the equivalent convolution and deconvolution algorithm.

Considerable attention has been paid in recent years to Monte Carlo (MC) methods owing to their competitive advantage in dealing with large systems [33]. The previous analytical methods mentioned become computationally difficult to apply to large systems because of the large number of events that must be evaluated grow on the order of m^n with increased system size. The MC method is presented after the Probability Tree Convolution Approach (PTC) to facilitate the explanation.

Another technique that has been brought forward interesting research is the Bloom method. This method includes the availabilities as integer programming similar to Balas's Branch and Bound method.

This work introduces the PTC tree approach first to explain the methods and secondly as a graphical approach to solving this problem [16]. A decision analysis based method is PTC. The PTC production costing method involves testing every possible state in a system. This can be prohibitively time-consuming for large systems.

Before proceeding, a list of assumptions is made in order to understand the limitations that are present with the application of each algorithm. The major assumptions for some methods include:

- Production costing is performed for each unit block segment.
- Minimum up/down times have been satisfied without explicit modeling.
- Costs related to unit outages include only the replacement generation energy.
- Outages are assumed to be independent events.
- Units have a two-state representation, i.e., either it is on or off.
- Derated states may be included as needed for increased accuracy.
- Probability of outage is given by the unit's forced outage rate (q).
- Segmented commitments of units follow only the economic demand order.
- Priority list method of committing units is based on the average production cost.
- When available capacity cannot satisfy the demand, assistance can be received from the capacity surplus generation or interchange, at a fixed cost.
- Amount of assistance is not constrained and is always readily available.

Without these assumptions, many of the techniques would be impractical. The removal of all assumptions is needed to provide a tool to accurately represent industry procedures.

4.1.11.1 Conventional Recursive Convolution Method

Reference [41] states that the EDDC is the most important concept established in the development of probabilistic production simulation technology. It integrates a generating unit's random outage with the random demand model.

Probabilistic reliability indices can be calculated by using the EDDC (referred to in the following as P), as given in [33].

$$P_i = \left(1 - \sum_{j=1}^{\mathrm{NS}} q_{ij}\right) P_{i-1}(x) + \sum_{j=1}^{\mathrm{NS}} q_{ij} P_{i-1}\left(x - C_{ij}\right) \tag{4.11}$$

where P_i equivalent demand duration curve (EDDC), x random variable of P, NS the total number of states, F_{oi} The outage capacity pdf of generator i, q_{ij} forced outage rate (FOR) of generator i at state j, C_{ij} outage capacity of generator i at state j.

The basic reliability evaluation indices (Loss of Demand Probability (LODP), Expected Energy Not Supplied (EENS), Energy Index of Reliability (EIR), Probabilistic production energy (E_i), Production cost (PC_i), capacity factor (CF_i), and total CO_2 emissions (TCO_2), are found using the effective demand duration curve, $P(x)$.

$$\mathrm{LODP} = \underset{\mathrm{IC}+D_{\mathrm{p}}}{P_{\mathrm{NG}i}}\,(x)\big|_{x=\mathrm{IC}}\ (\mathrm{h/year}) \tag{4.12}$$

$$\mathrm{EENS} = \sum_{x=\mathrm{IC}}^{\mathrm{IC}+D_{\mathrm{p}}} \Phi_{\mathrm{NG}}(x)\,(\mathrm{MWh/year}) \tag{4.13}$$

$$\mathrm{EIR} = \left(1 - \frac{\mathrm{EENS}}{\mathrm{ED}}\right)(\mathrm{pu}) \tag{4.14}$$

$$\Delta E_i = \mathrm{EENS}_{i+1} - \mathrm{EENS}_i\,(\mathrm{MWh/year}) \tag{4.15}$$

$$\Delta PC_i = F_i(\Delta E_i, \mathrm{LODP}_{i-1})\ (\$) \tag{4.16}$$

$$\mathrm{CF}_i = (\Delta E_i/\mathrm{CAP}_i/T) \times 100 \tag{4.17}$$

$$\mathrm{TCO}_2 = \sum^{\mathrm{NT}} \sum^{\mathrm{NG}} (\varepsilon_i \Delta E_{\mathrm{in}})\ (\mathrm{ton/year}) \tag{4.18}$$

where L_{p} peak demand [MW], IC_i installed capacity of generator i[MW], ED total demand energy [MWh], NG the total generator number, P_{NG} the final effective demand duration curve, ξ CO_2 emission coefficient of the ith unit [Ton/MWh].

4.1.11.2 Draw Backs of the Analytical Method

It is worth noting that the analytical method, the conventional recursive convolution method, is efficient in computation in terms of computer time. However, it has a major flaw in recognizing the chronological variations of power generation and demand. Furthermore, there is no direct representation of minimum run time, minimum down time, or even the role of start-up costs and spinning reserve in commitment decisions.

Though here in this study no minimum run/down times or start-up costs are considered, there might be chronological variations of power generations. This case might arise upon demanding the last unit; when the needed generation left is below the unit's minimum generation, the unit is turned on to its minimum and the unit before is decreased to satisfy reserve requirements at minimum cost.

Otherwise if there are no chronological variations of power generations, both the conventional recursive convolution (CRC) and the state enumeration methods (SE) should give the same results.

4.1.12 Probability Tree Convolution Method

Decision analysis [55–57] is a systematic approach for decision making which enables solution of problems where uncertainty is a prominent factor. A normative model is developed to form the decision problem statement, enable logical analysis, and focus on the production of a recommended course of action. This methodology is most useful in managerial situations including significant risk. The resulting process when documented is capable of generating optimal strategies for multi-stage decision problems under a multiplicity of contingencies.

The position held in this work is that formal decision modeling contributes positively to decision making by cross-checking and evaluating outcomes with a logically robust model. Data analysis and experience contribute to the formal modeling process with insights and knowledge that are incorporated into the models. The resulting data analysis yields better results than the individual parts by themselves. Modeling produces a graphical representation of mental conceptions useful for organizational communication. The model can serve as a plan of action as well as an instrument for project management.

4.1.12.1 Decision Trees

An example traditional tree diagram is shown in the following. A square box traditionally shows a decision node, a round box a natural node, and a triangle the leaves for the value given the decisions made and the unknown events have occurred. We use rounded boxes in this work to denote probabilities from statistical

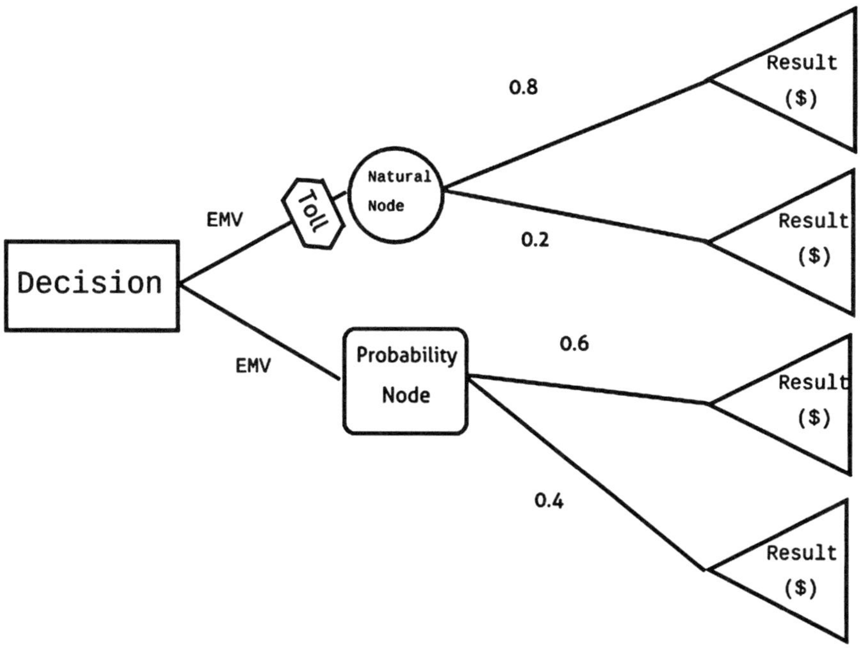

Fig. 4.17 Decision tree

analysis. The toll for a decision is represented by a diamond ended box. The toll can also be combined with the expected profit at the end of the branch, the tree node. Probability trees are extensions of this simple example. Natural nodes may be added after the first natural node (added to the right) for multiple uncertain events, even if correlated. Decision nodes may be added to show subsequent decisions. The initial model should always include the "do nothing" alternative.

A decision problem outcome depends on the action alternative selected by the decision maker and the state of nature which occurs. Outcomes may be favorable or unfavorable depending on the events that do happen. The payoff matrix is another way of showing the value of a set of possible outcomes (Fig. 4.17).

States of nature, being uncertain, require probabilities to denote the likelihood of occurrence. It is not unusual for subjective probabilities to be used when statistical data is not available. However, statistical values result in more accurate assessments.

The convolution of production costing probabilities may be shown as a probability tree with only natural or statistical nodes. This is a probability tree that is evaluated in the same fashion as the generation first method with generation convolved first, and then demand second [16].

Convolution can also be applied by use of the probability tree discussed above. An example tree is shown in the following figure, assuming that two states are used for each generator. The demand curve at each leaf is either the cumulative demand

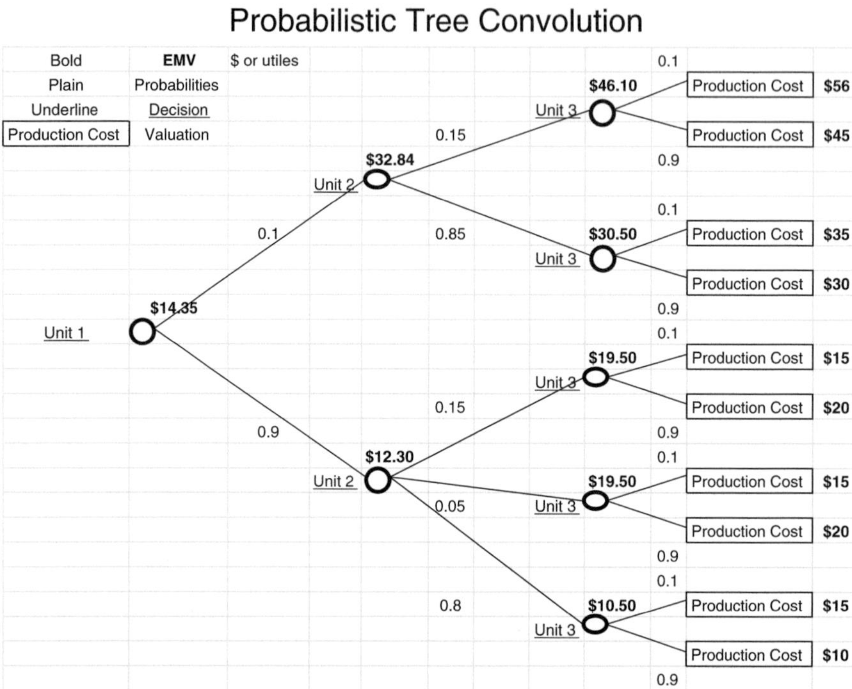

Fig. 4.18 Probabilistic tree convolution

curve or the hourly demand curve. Each leaf is solved by the appropriate scheduling and/or dispatching algorithm. The hourly duration curve is solved by unit commitment or by security constrained unit commitment. The transmission system is solved for the most accurate values to include overloads and transmission contingencies. The production cost is found by Economic Dispatch, Unit Commitment, or Security Constrained Unit Commitment (SCUC) over the period of interest. The algorithms used depend on the accuracy required, the constraints to be modeled, and the demand forecast model. The equivalent demand duration curve is calculated by accumulating the results from the unit commitment and/or transmission contingency analysis algorithm (Fig. 4.18).

The number of paths demonstrates the curse of dimensionality, even for such a small system as this one. The paths are shown in the following Table 4.2.

We use a table approach since the calculations for this work were performed in Excel (Table 4.3).

We use the font encodings to show each step of the process (Table 4.4).

Table 4.2 Lists of paths and event valuations at the leaves

Unit 1	Unit 2	Unit 3	Path	Value
0.9	0.8	0.9	1	$56.00
0.9	0.8	0.1	2	$45.00
0.9	0.05	0.9	3	$34.00
0.9	0.05	0.1	4	$30.00
0.9	0.15	0.9	5	$15.00
0.9	0.15	0.1	6	$20.00
0.1	0.15	0.1	7	$15.00
0.1	0.15	0.9	8	$20.00
0.1	0.15	0.1	9	$15.00
0.1	0.15	0.9	10	$10.00

Table 4.3 EMV calculations

EMV	Unit 1	EMV	Unit 2	EMV	Unit 3	Cost
$14.35	0.1	$32.84	0.15	$46.10	0.1	$56.00
					0.9	$45.00
			0.85	$30.50	0.1	$35.00
					0.9	$30.00
	0.9	$12.30	0.15	$19.50	0.1	$15.00
					0.9	$20.00
			0.05	$19.50	0.1	$15.00
					0.9	$20.00
			0.8	$10.50	0.1	$15.00
					0.9	$10.00

Table 4.4 Table notation

Item	Value	Notation
EMV	$ or Utiles	Bold
Probabilities	Normalized (sum to 1.0)	Plain
Decision	$ or Utiles	Underline

4.1.12.2 Expected Monetary Value Models (EMV & EOL)

Once a probability distribution has been assessed for each set of uncertain states of nature—and this can always be done, subjectively—it is straightforward to apply the next step: compute the expected value for each action alternative. Since there are two ways to look at the same problem (actual monetary values and opportunity losses), we can compute the expected values on either one of the payoff tables. The EMV is an indicator of the information captured by the process. The EMV will slowly change after the most significant paths are included in the convolution process.

Table 4.5 Payoff matrix

	State			Payoff	
Action	H	M	L		
p_j's	0.1	0.6	0.3	**EMV**	
Base unit	15	3	−6	**1.5**	
Combined cycle unit	9	4	−2	**2.7**	<=max EMV = EMV*
Peaking unit	3	2	1	**1.8**	

Table 4.6 Payoff list

Optimal	Decision	**EMV**	State	Probability	Value
	Base unit	1.5	H	0.1	15
			M	0.6	3
			W	0.3	−6
=>	Combined cycle unit	2.7	H	0.1	9
			M	0.6	4
			W	0.3	−2
	Peaking unit	1.8	H	0.1	3
			M	0.6	2
			W	0.3	1

Expected Monetary Value

Referring to the original payoff matrix, the formula for expected monetary value (EMV) is:

$$\mathrm{EMV}(A_i) = E(A_i) = \sum_j p_j\left(R_{ij}\right) \tag{4.19}$$

where i refers to the matrix's rows and j refers to the columns. Thus, using the probability distribution derived previously, we obtain the following (Table 4.5).

Alternatively, we can make use of the decision tree matrix in Excel format (Table 4.6):

Expected Opportunity Loss

One of the important market indicators is the Expected Opportunity Loss (EOL). Recall from the Savage criterion that an opportunity loss is the payoff difference between the best possible outcome under S_j and the actual outcome resulting from choosing A_i given that S_j occurs. Referring now to the opportunity loss matrix, the formula for expected opportunity loss (EOL) is:

Table 4.7 Minimum EOL

A	S				
	H	M	L		
Probabilities	0.1	0.6	0.3	**EMV**	
Series 1	0	1	7	2.7	
Series 2	6	0	3	1.5	<=min EOL = EOL*
Series 3	12	2	0	2.4	

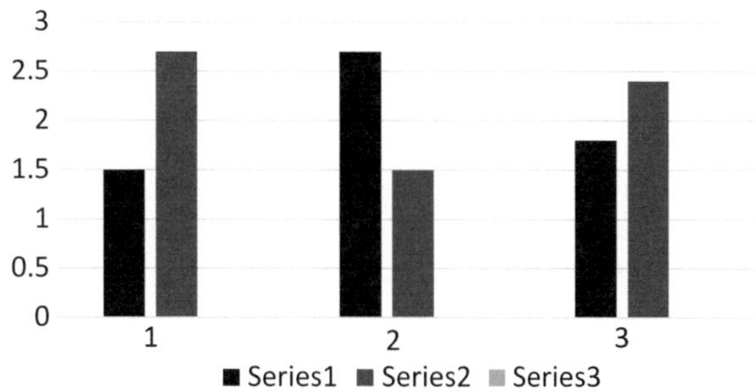

Fig. 4.19 Relationship between EMV and EOL

Table 4.8 EVgPI calculation

EVgPI	State	Probability	Value
4.2	H	0.1	15
	M	0.6	4
	L	0.3	1

$$\text{EOL}\ (A_i) = E\ (A_i)\ =\ \Sigma_j p_j(\ \text{OL}_{ij}) \tag{4.20}$$

Obviously, the same probability distribution applies (since the states of nature are the same) (Table 4.7):

EOL can also be depicted with a decision tree, of course. (Exercise left to the reader.)

Note that for a given probability distribution, the expected payoffs (EMV and EOL) for every action alternative A_i always add up to a constant. In our case, they always add up to 4.2. Thus, the Max EMV corresponds with the Min EOL (Fig. 4.19).

The Expected Value given Perfect Information (EVgPI) is 4.2, and is obtained as follows (Table 4.8).

We obtain the expected value of the above lottery (the EVgPI) thusly:

$$\text{EVgPI}\ =\ \Sigma_j p_j(R_{ij}*) \tag{4.21}$$

The Expected Value of Perfect Information (EVPI) Is then:

$$\text{EVPI} = \text{EVgPI} - \text{EMV}^* \tag{4.22}$$

If the above lottery is solved using opportunity losses instead of monetary values, we find:

$$\text{EVPI} = \text{EOL}^* \tag{4.23}$$

4.1.12.3 State Enumeration Method

In planning generation supply for the next hour and thus the expected cost of production, the system operator is basically acting on the basis of expected monetary value (EMV). In general, one obtains the EMV of a "gamble" with several possible outcomes by multiplying each possible cash outcome by its probability and summing these products over all the possible outcomes [58].

For example, if an EMV'er is given the chance at a gamble with payoffs of $0.00 or $100.00 with equal probability of 0.5, his EMV would be:

$$\text{EMV} = (0.5)(\$0.00) + (0.5)(\$100.00) = \$50.00 \tag{4.24}$$

Thus this gamble is worth $50.00. As a result, given different projects or options to choose from, the EMV'er would choose the alternative with the most expected profit or the least expected cost.

Tying this with production costing estimation, the structured tree would enumerate all the possible generator status combinations. The first node, as shown in the Fig. 4.16, would be the first unit on the priority list with the least production cost from which there are two possible state outcomes either it is out of service, s1 = 0, or in service, s1 = 1.

The next stage of the structured tree is the status of generator 2 with 2 possible outcomes. This process is repeated until all possible states are counted for. If on one branch the demand is satisfied by generator i, then the rest of the $n - i$ branches are trimmed.

With the structured tree built, at most there would be 2^N possible state combination, status = (s1, s2, s3, . . ., si, . . ., sn), and their corresponding probabilities (p, q).

Now for each possible status k, economic dispatch is performed on the available units. The economic dispatch is represented as:

$$z_k = \text{Min}X(\text{incost}_i \times p\text{gen}_i) + \text{inscost} \times \text{insdemand} \tag{4.25}$$

subject to:

$$X p\text{gen}_i + \text{insdemand} = \text{demand} \tag{4.26}$$

$$p\text{min}_i < p\text{gen}_i < p\text{max}_i \tag{4.27}$$

Table 4.9 Generator states, generation, probability, and EMV's

State	S3	S2	S1	pgen3 (MW)	pgen2 (MW)	pgem (MW)	Insurance (MW)	Prob	**EMV** ($)
1	1	1	1	100	550	400	0	0.648	57,646
2	1	1	0	400	600	0	50	0.072	822.82
3	1	0	1	400	0	400	250	0.072	389.02
4	1	0	0	400	0	0	650	0.008	289.23
5	0	1	1	0	600	400	50	0.162	767.74
6	0	1	0	0	600	0	450	0.018	499.93
7	0	0	1	0	0	400	650	0.018	641.48
8	0	0	0	0	0	0	1,050	0.002	105.00

where z_k = system operating cost of status combination k, ($), incost_i = generation cost of unit i, $/MWh, pgen_i = generation of unit i, MW, demand = energy demand, MW.

This problem could be solved by inspection: demand the generators by increasing operating cost until demand is met. As a result, the expected monetary value of each possible combination, 1, 2, . . ., 2^N, is:

$$\text{emv}_k = (z_k + \text{status}_k \times \text{fixcost}) \times \text{prob}_k \tag{4.28}$$

where emv_k = expected monetary value of status combination k, ($), z_k = system operating cost of status combination k, ($), status_k = row matrix of status combination k, fixcost = column matrix of fix cost, ($), prob_k = probability of status combination k.

The expected production cost, epc, of the structured tree is the sum of the emv_k's.

4.1.12.4 State Enumeration Method Example

The following gives the state enumeration, their probabilities, the corresponding generations of the units, and the MW insurance needed to satisfy the demand. As can be seen from the table, in state 1 unit three is operated at its minimum and unit two at below maximum generation—a chronological variations of power generation (Table 4.9).

From these data, the expected generation and capacity factor is obtained for each unit. Note that the capacity factor for each unit is obtained by summing the probabilities corresponding to when each unit is on. For example unit one has a capacity factor, Capf1, (0.648 + 0.072 + 0.072 + 0.008 = 0.8). This matches the result obtained in the recursive convolution method (CRC). The expected operation cost of the production is calculated as $11,279.84. Compared to the recursive convolution method, the error is minimal in this very simple case.

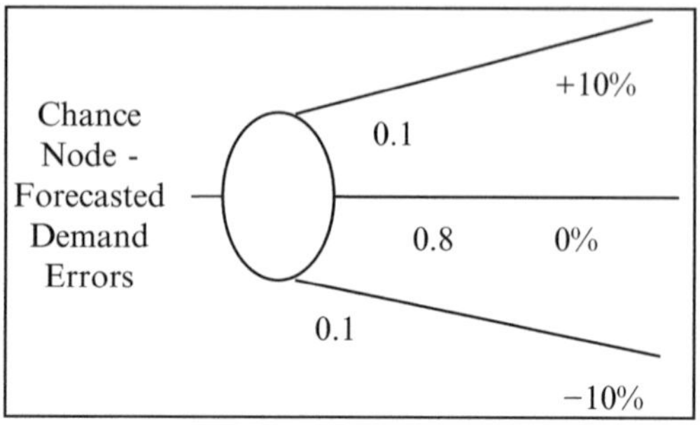

Fig. 4.20 Forecast error model

4.1.12.5 Forecast Error

The forecast error for demand, for wind, or for solar can be included simply my adding another node with probabilistic links to the expected forecast and the first or second moments. This is similar to the method of aggregating the wind generator to three levels of generation in the above. There is no need to limit the probabilities to three levels except that high, medium, and low forecasts are rather natural. Indeed, any number of links can be added as is desired to increase the accuracy. However, one should employ the parsimonious concept to model what is an impact instead of simply increasing links to "make sure" (Fig. 4.20).

This same technique can be used to find the wind generation distribution instead of the linearization described above. The solar distribution impact on production is another application of this approach.

4.1.12.6 Linear Programming Based Production Cost Model

The unit and contract costs as solved by EDC, UC, or SCUC to find the leaf values. This is equivalent to the unit costs found by integrating the EDDC between the unit segments low capacity level and the high capacity for each segment.

One approach is based on LP based production costing algorithm for multiple time periods. For each path, we form a linear program to forecast the production costs. The probabilistic production costs can be computed by calculating expected production cost using the decision tree. The advantage of this approach is that after the first leaf is evaluated, each subsequent leaf can be found by parametric linear programming.

Problem formulation: The costs associated with satisfying the power production requirements expressed by a demand duration curve are:

1. Start-up cost.
2. Unit running costs.

The constraints can be listed as:

1. Enough units should be committed to satisfy the ancillary requirements.
2. Enough power should be dispatched to satisfy the demand requirements.
3. Only resources available can participate in (1) or (2).
4. Resources for ancillary services do not have to be units.

Mathematically, the problem can be expressed as follows:

$$T_{\mathrm{p}} + G_{\mathrm{a}} + I_{\mathrm{p}} \geq L \text{ (Demand constraint)} \tag{4.29}$$

$$T_{\mathrm{f}} = H(G_{\mathrm{a}}, T_{\mathrm{p}}, I_{\mathrm{p}}) \text{ (Conversion curve)} \tag{4.30}$$

$$T_{\mathrm{f}} \leq T_{\mathrm{cmax}} \text{ (Transmission capacity constraint)} \tag{4.31}$$

where H = nonlinear relationship between power produced and purchased, $F(p)$ = production cost of the utility company, I = cost of insurance, T_{p} = transaction power, G_{a} = summation of maximum capacity of available generating units, I_{p} = power purchased from insurance company, L = total demand of the utility company, T_{f} = demanding of transmission line for transaction, T_{cmax} = capacity of transmission line for transaction.

Probabilistic modeling of the abovementioned random variables can be done using power system reliability data. Probability of demanding transmission line exceeding its capacity can be achieved using probabilistic demand flow formulation. The cost calculation can be done by identifying a suitable production costing model.

Minimize

$$\sum_{u=1}^{u\max}\sum_{s=1}^{s\max}(\mathrm{crun}(u,s)\mathrm{xrun}(s) + \mathrm{cprod}(u,s)\mathrm{xprod}(u,s) \tag{4.32}$$

Subject to

$$\sum_{u=1}^{u\max}(\mathrm{capmax}(u) \times \mathrm{xrun}(i)) \geq \mathrm{spin}(s) \tag{4.33}$$

$$\sum_{u=1}^{u\max}\sum_{i=s}^{s\max}\mathrm{capmin}(u)\mathrm{xrun}(i) + [\mathrm{capmax}(u) - \mathrm{capmin}(u)]\mathrm{xprod}(u,s) \geq \mathrm{demand}(s) \tag{4.34}$$

$$\mathrm{xprod}(u,s) - \sum_{i=s}^{s\max}\mathrm{xrun}(i) \leq 0 \tag{4.35}$$

$$\sum_{s=1}^{s\max}\mathrm{xrun}(s) \leq u\max \tag{4.36}$$

where xrun(s) = number of units running through segment s, xprod(u,s) = level of output above minimum of unit u in segment s, capmin(u) = minimum capacity of u, capmax(u) = maximum capacity of u, crun(u,s) = cost of runing unit u through segment s at minimum output, cprod(u,s) = costs above crun of u at full demand in segment s, spin(s) = spinning MW requirement for segment s, demand(s) = Generation MW requirement in segment s.

4.1.13 Bloom Gallant Linear Programming

From the buyers' perspective, the transaction selection problem can be defined as follows. Given a set of possible transactions, select a transaction plan that minimizes the net expected cost of production and cost of insurance, which are subject to risks due to uncertainty in power system reliability. Mathematically, the problem can be expressed as follows for each period (K):

Minimize Expected cost of production and insurance over all periods:

$$\sum_{k=1}^{K}\left[\sum_{i=1}^{I}[F(i,k) + ((\text{EENS}) \times EIE(k))]\right] \tag{4.37}$$

Subject to

$$\sum_{k=1}^{K}\sum_{i=1}^{I} D_{i,m}^{k} e_i^k = F^m; \quad m = 1, \ldots, M \tag{4.38}$$

$$\sum_{i=1}^{I} A_{i,j}^{k} e_i^k = B_j^k; \quad j = 1, \ldots, J^k; \quad k = 1, \ldots, K \tag{4.39}$$

$$\sum_{i \in \Omega} e_i^k \leq W^k(\varnothing) - W^k(\Omega); \forall \Omega \subset \{1, \ldots, I\}; \quad k = 1, \ldots, K \tag{4.40}$$

$$\sum_{i=1}^{I} e_i^k = W^k(\varnothing); \quad k = 1, \ldots, K \tag{4.41}$$

$$e_i^k \geq 0; \quad i = 1, \ldots, I; \quad k = 1, \ldots, K \tag{4.42}$$

The first equation is the inter period constraints which couple the production of a set of plants over different time periods, especially useful for hydro, take or pay fuel contracts, etc. The second equation are the intra-period constraints which couple production of a set of plants within a given time period, especially environmental and dispatch limits. Dispatch limits include energy storage plants, must-run plants, multiple block plants, two area systems with transmission limits, limited energy plants, as well as dispatchable plants. The third equation and the fourth equation are the facet constraints [4] used to express probabilistic production costing as a linear programming problem.

The difficulty with Bloom-Galant's method is the same as Balas's 0-1 algorithm. The curse of dimensionality and the inability to generate a plane on an integer solution is yet to be solved.

4.1.14 Monte Carlo

The Monte Carlo process is a sampling procedure based on system simulation. Various random samples from an infinite population, *X*, having the same size are taken; the mean of these samples would form another distribution, called the sampling distribution, which tends to be a normal distribution even if the underlying parent distribution is not as the sampling size gets larger according to the Central Limit Theorem (CLT).

Central Limit Theorem: If $X1, \ldots, Xt$ are independent and identically distributed random variables with mean band variance s_2 and are sampled at *T* different time values:

$$X = (X_1 + X_2 + \ldots + X_t)/T \tag{4.43}$$

then

$$Z = [X - E(X)] /([\text{var}(X)]1/2) \tag{4.44}$$

has a probability distribution that approaches the standard normal, *N*(0,1), as *T* approaches infinity.

The Monte Carlo method has been the preferred solution for almost two decades [5, 28, 40, 50, 52, 58–63].

One method of the Monte Carlo simulation method is called the brute force Monte Carlo Method [60]. In this approach an "experiment" is conducted to build a sample of *T* randomly chosen distinct system states and the status. Calculations are carried out as done in the state enumeration method to obtain the EMV_i's ($i = 1, 2,\ldots,T$); the sample's estimated production cost, EPC, for this experiment, *k*, is the mean of the EMV's. This process of running an experiment is repeated *k* times until "convergence" to a normal distribution is obtained; the mean of the experiments would give the expected production cost.

This method according to Parker [28] needs a large sample size to obtain a reasonable estimation of the expected production costing method. Thus the Monte Carlo method is used with Latin Hypercube sampling to "more accurately capture uncertainties with limited draws." Parker suggests to stratify the outage states into equal intervals on the cumulative probability scale.

Huang and Chen [63] present another approach of stratifying a population using what is called the "cum sqrt(mu f)" rule. In most programs, each system state is generated using a random number generator which produces uniformly distributed random numbers chosen from a uniform distribution on the interval (0.0, 1.0). Thus each system status, $S = (s1, s2, s3, sn)$, of the sample is generated at random. The corresponding generator status, *si*, given by:

$$\begin{aligned} &s_{i,t} = 0 \text{ if } R < \text{forate}_i \text{ from Random Number generator (outaged),} \\ &\text{else} \\ &s_{i,t} = 1 \text{ if } R > \text{ forate}_i(\text{available}) \end{aligned} \tag{4.45}$$

According to Billinton [17, 26], for such distributions the size of the samples necessary to estimate a mean varies as the square of a coefficient of variation of the sampling distribution. This coefficient which is the convergence criterion is the ratio of the standard deviation of the sample mean *E* (epc) of the system epc over the sample mean *E* (epc). The simulation is terminated when:

$$\text{Ratio} = a[E(\text{epc})]/E(\text{epc}) < e \tag{4.46}$$

where e is the maximum allowable error. When a smaller e is specified, more iterations are needed. Note that a [E(epc)] is statistically expressed as:

$$S[E(\text{epc})] = s(\text{epc})/V(k) \tag{4.47}$$

where s (epc) is the standard deviation of the k repeated experiments.

In summary, this procedure is summarized in the following pseudo-code (Table 4.10).

The sequential Monte Carlo simulation (MCS) method is a powerful tool for power systems adequacy assessment. Sequentially sampling the states duration, this method can incorporate the stochastic behavior of the system components, time-dependent issues like the renewable power production, reservoir operating rules, scheduled maintenance, complex fuel contracts, etc. It provides the probability distribution of the reliability indices. Despite these advantages, the simulation time of the sequential MCS method is its major weakness. The objectives of this work is to present algorithmic advances that can improve sequential MCS method as applied to the adequacy assessment of the generating capacity and composite (generation and transmission) system.

Table 4.10 Monte Carlo Pseudo-code

select number of samples and epsilon, k = 1, sample size = T	
While ratio > epsilon k = k + 1 do	
	generate a sample of system states of size T
	calculate the emvi for each of the T samples
	calculate the mean of the emvT 's = epck
	calculate the stopping criterion ratio
	Endwhile
stop the estimated production cost is the mean of the epck's	

Recent advances have explored the application of the Cross-Entropy (CE) method and the Importance Sampling (IS) variance reduction technique in the sequential MCS method [64]. A new algorithm was proposed to calculate the CE-optimal IS distribution for the generating capacity adequacy assessment. This new CE-based algorithm steams from the mathematical analysis of the CE equations that has demonstrated that the CE-optimal IS distribution can be obtained by simply dividing the annualized reliability indices for different configurations of the generating system. This algorithm, whose core is the fast Fourier transform, is equivalent to the standard CE optimization algorithm in accuracy and computational effort. The relevant feature of the new CE-based algorithm when compared to the standard CE optimization algorithm is its simplicity of implementation. Several strategies for modeling the generating units with time-dependent capacity in the CE-based algorithms were also suggested and their impact on the simulation time duly analyzed. The second part of this advance also proposed and examined a CE optimization algorithm for the composite system adequacy assessment.

Another part of this improvement introduced the innovative application of a Population-Based Method (PBM) to improve the efficiency of the sequential MCS method. "The proposed methodology consists of two phases. Firstly, a list of high probability states that cannot supply the peak load is created by a PBM. The PBM used takes advantage of the space-covering characteristics of the Evolutionary Particle Swarm Optimization (EPSO) metaheuristic. Secondly, the states sampled by the sequential MCS method are compared to those on the list to decide whether full evaluation should be performed or not. If a state proceeds to evaluation, the yearly load model, the time-dependency of the capacity of the generating units, and other chronological features are sequentially followed to form system states. These system states may or may not have loss of load. If the state sampled is not in the list, then it is assumed that no loss of load occurs throughout its duration" [62].

The results obtained from using the CE method and IS in the sequential MCS method reported remarkable acceleration in the estimation of the reliability indices for the generating capacity and composite system (it was reported that in some experiments, the time gain over the crude sequential MCS method is more than 60 times). Moreover, it was observed that the acceleration increased as the system become more reliable. Unfortunately, the sequential MCS method cannot provide accurate probability distributions for the reliability indices if the CE method and IS are used. On the other hand, the experiments carried out in the third part CE method and IS were used, the experiments carried out demonstrated that the accelerations achieved are only comparable to the ones obtained by the CE method and IS if the system is unreliable. Despite this disadvantage, this methodology can obtain accurate probability distributions for the reliability indices if the classification process does not fail to detect the states that need evaluation. More advances in these areas are warranted.

4.2 Conclusion

The applications of PPC are expanding as markets are evolving. One such example is the capacity market to recover fixed costs. Also, the capacity credit for renewable energy has been investigated for capital budgeting comparisons [27, 65, 66]. The PPC methods are the central computational engine for market analysis, bidding, risk analysis, hydro thermal coordination, maintenance scheduling, contracts evaluation of interchange through demand response or demand side management. Clearly, the computational requirement of the core assessment algorithm determines the speed and accuracy of all subsequent analysis applications.

The computer time needed to solve the above is now within the reach of many companies, especially Independent Power Producers (IPPs), municipal utilities, and rural cooperatives. Distributed computation can be employed to solve the production cost models at each leaf concurrently. The cost of such computer networks is easily justified to assess and then to manage the risk within a short solution time. Concurrent processing should not be confused with parallel processing. The cost of workstations enables such algorithms as demonstrated by the proliferation of computation software, such as MATLAB or PYTHON, across multiple workstations as a simple toolbox.

References

1. Allen RN, Billinton R, Breipohl AM, Grigg CH (1999) Bibliography on the application of probability methods in power system reliability evaluation – 1992–1996. IEEE Trans Power Syst 14(1):51–57
2. Allen RN, Billinton R, Breipohl AM, Grigg CH (1994) Bibliography on the application of probability methods in power-system reliability evaluation – 1967–1991. IEEE Trans Power Syst 9(1):41–48
3. Billinton R, Fotuhi-Firuzabad M, Bertling L (2001) Bibliography on the application of probability methods in power system reliability evaluation 1996–1999. IEEE Trans Power Syst 16(4):595–602
4. Bloom JA (1992) Representing the production cost curve of a power system using the method of moments. IEEE Trans Power Syst 7:1370–1377
5. Carvalho L (2013) Advances on the sequential Monte Carlo reliability assessment of generation-transmission systems using cross-entropy and population-based methods. Thesis for Doctor of Philosophy, Universidade do Porto, Faculdade de Engenharia, Supervisor: Professor Vladimiro Henrique Barrosa Pinto de Miranda, Ph.D.; Co-supervisor: Mauro Augusto da Rosa, Ph.D.; Porto, Portugal
6. Giorsetto P, Utsurogi KF (1983) Development of a new procedure for reliability modeling of wind turbine generators. IEEE Trans Power Appar Syst 102(1):134–143
7. Liang R-H, Liao J-H (2007) A fuzzy-optimization approach for generation scheduling with wind and solar energy systems. IEEE Trans Power Syst 22(4):1665–1674
8. Marwali MKC, Ma H, Shahidehpour SM, Abdul-Rahman KH (1998) Short-term generation scheduling in photovoltaic-utility grid with battery storage. IEEE Trans Power Syst 13(3):1057–1062

9. Pereira V, Gorenstin BG, Fo M (1992) Chronological probabilistic production costing and wheeling calculations with transmission network modeling. IEEE Trans Power Syst 7(2):885–891
10. Tome Saraiva J, Miranda V, Pinto LMVG (1996) Generation/transmission power system reliability evaluation by Monte-Carlo simulation assuming a fuzzy load description. IEEE Trans Power Syst 11(2):690–695
11. Schilling MT, Billington R, Leite da Silva AM, El-Kady MA (1989) Bibliography on composite system reliability. IEEE Trans Power Syst 4(3):1122–1132
12. Whitt W (1992) Asymptotic formulas for Markov processes with applications to simulation. Oper Res 40:279–291
13. Wu FF, Tsai YK (1983) Probabilistic dynamic security assessment of power systems: part 1 – basic model. IEEE Trans CAS 148–149
14. Saintcross J, Piwko R, Bai X, Clark K, Jordan G, Miller N, Zimberlin J (2005) The effects of integrating wind power on transmission system planning, reliability, and operations, report on phase 2: system performance evaluation. GE Energy Consulting, prepared for The New York State Energy Research And Development Authority
15. Day JT (1971) Forecasting minimum production costs with linear programming. IEEE Trans Power Appar Syst 90(2):814–823
16. Wood A, Wollenberg B, Sheblé G (2014) Power generation operation and control, 3rd edn. Wiley, New York, NY
17. Billinton R (1970) Power system reliability evaluation. Gordon and Breach, New York, pp 97–102
18. Endrenyi J (1978) Reliability modeling in electric power systems. Wiley, New York
19. Ross SM (1993) Introduction to probability models, 5th edn. Academic, New York
20. Singh C, Lago-Gonzalez A (1985) Reliability modeling of generation systems including unconventional energy sources. IEEE Trans Power Appar Syst 104(5):1048–1056
21. Wang X, McDonald JR (1994) Modern power system planning. McGraw-Hill, London
22. Wang H, Dai T, Thomas RJ (1984) Reliability modeling of large wind farms and associated electric utility interface systems. IEEE Trans Power Appar Syst 103(3)
23. Wood A, Wollenberg B (1996) Power generation operation and control, 2nd edn. Wiley, New York, NY
24. Garver LL (1972) Adjusting maintenance schedules to levelize risk. IEEE Trans Power Appar Syst 91:2057–2063
25. Baleriaux H, Jamoulle E, Linard De Guertechin F (1967) Simulation de l'exploitation d'un parc de machines thermiques de production d'electricite couple a des stations de Pompage. Revue E (Edition SRBE) 5:225–245
26. Billinton R, Ringlee RJ, Wood AJ (1975) Power system reliability calculations. Oper Res 23(1):182–184
27. Delson JK, Feng X, Smith WC (1991) A validation process for probabilistic production costing programs. IEEE Trans Power Syst 6(3):1326–1336
28. Parker C, Stremel J (1996) A smart Monte Carlo procedure for production costing and uncertainty analysis. Proc Am Power Conf 58(II):897–900
29. Ryan SM, Mazumdar M (1990) "Effect of frequency and duration of generating unit outages on distribution of system production costs. IEEE Trans Power Syst 5:191–197
30. Ru RS, Toy P, Schenk KF (1980) Expected energy production cost by methods of moments. IEEE Trans Power Appar Syst 1917–1980
31. Wang L (1989) Approximate confidence bounds on Monte Carlo simulation results for energy production. IEEE Trans Power Syst 4(1):69–74
32. Breipohl AM, Lee FN, Zhai D, Adapa R (1992) A Gauss-Markov load model for application in risk evaluation and production simulation. IEEE Trans Power Syst 7:1493–1499
33. Lin M, Breipohl A, Lee F (1989) Comparison of probabilistic production cost simulation methods. IEEE Trans Power Syst 4(4):1326–1333

34. Mazumdar M, Yin CK (1989) Variance of power generating system production costs. IEEE Trans Power Syst 4:662–667
35. Ryan SM, Mazumdar M (1992) Chronological influences on the variance of electric power production costs. Oper Res (Suppl 2):S284–S292
36. Sager MA, Ringlee RJ, Wood AJ (1972) A new generation production cost program to recognize forced outages. IEEE Trans Power Appar Syst 91:2114–2124
37. Valenzuela J, Mazumdar M (2000) Statistical analysis of electric power production costs. IIE Trans 32:1139–1148
38. Grigg C, Wong P, Albrecht P, Allan R, Bhavaraju M, Billinton R, Chen Q, Fong C, Haddad S, Kuruganty S, Li W, Mukerji R, Patton D, Rau N, Reppen D, Schneider A, Shahidehpour M, Singh C, The IEEE Reliability Test System-1996 (1999) A report prepared by the reliability test system task force of the application of probability methods subcommittee. IEEE Trans Power Syst 14(3):1010–1020
39. Park J, Liang W, Choi J, El-Keib AA, Shahidehpour M, Billinton R (2009) Probabilistic reliability evaluation of power system including solar/photovoltaic cell generator. IEEE PES GM2009, Calgary, AB, Canada
40. Carvalho L, Gonzalez-Fernandez RA, Leite da Silva AM, da Rosa MA, Miranda V (2013) Simplified cross-entropy based approach for generating capacity reliability assessment. IEEE Trans Power Syst 28(2):1609–1616
41. Sheble GB (1999) Computational auction mechanisms for restructured power industry operation. Springer, New York, NY
42. Sheble GB (1999) Decision analysis tools for GENCO dispatchers. IEEE Trans Power Syst 14(2):745–750
43. Billinton R, Gan L (2000) Wind power modeling and application in generating adequacy assessment. Proceedings, the 14th power systems computation conference, Sevilla, Spain
44. Billinton R, Chowdhury AA (1992) Incorporation of wind energy conversion systems in conventional generating capacity adequacy assessment. IEE Proceedings-C 139(1):47–55
45. Billinton R, Chen H, Ghajar R (1996) A sequential simulation technique for adequacy evaluation of generating systems including wind energy. In: IEEE/PES 1996 Winter Meeting, vol 96, Baltimore, MD, WM 044-8 EC
46. Carvalho L, Issicaba D, da Rosa MA, Ramos JPV, Miranda V, Leite da Silva AM (2012) Probabilistic analysis for maximizing the grid integration of wind power generation. IEEE Trans Power Syst 27(4):2323–2331
47. Gavanidou ES, Bakirtzis AG, Dokopoulos PS (1992) A probabilistic method for the evaluation of the performance and the reliability of wind diesel energy systems. IEEE Paper no. 92 SM 526-6 EC
48. Lee FN, Lin M, Breipohl AM (1990) "Evaluation of the variance of production cost using a stochastic outage capacity model. IEEE Trans Power Syst 5:1061–1067
49. Michaelides JM, Votsis PP (1991) Energy analysis and solar energy development in Cyprus. Comput Contr Eng J 2(5):211–215
50. Sullivan RL (1977) Power system planning. McGraw-Hill, New York
51. Zaininger HW (Power Technologies, Inc) (1977) Synthetic electric utility systems for evaluating advanced technologies. Electric Power Research Institute
52. Billinton R, Li W (1991) Hybrid approach for reliability evaluation of composite generation and transmission systems using Monte-Carlo simulation and enumeration technique. IEE Proc C 138(3):233–241
53. Karki R, Hu P, Billinton R (2006) A simplified wind power generation model for reliability evaluation. IEEE Trans Energ Convers 21(2):533–540
54. Mazumdar M, Bloom JA (1996) Derivation of the Baleriaux formula of expected production costs based on chronological load considerations. Electr Power Energ Syst 18:33–36
55. IEEE Reliability Test System (1979) IEEE Trans Power Appar Syst 98:2047–2054
56. Raiffa H (1968) Decision analysis. Addison-Wesley, Reading, MA

57. Sheble GB (1989) Real-time economic dispatch and reserve allocation using merit order loading and linear programming rules. IEEE Trans Power Syst 4(4):1414–1420
58. Billinton R, Gan L (1993) Monte Carlo simulation model for multiarea generation system reliability studies. IEE Proc C 140(6):532–538
59. Billinton R, Li W (1992) A Monte Carlo method for multi-area generation system reliability assessment. IEEE Trans Power Syst 7(4):1487–1492
60. Billinton R, Lian G (1991) Monte Carlo approach to substation reliability evaluation. IEE Proc C 140(2):147–152
61. Billinton R, Li W (1994) Reliability assessment of electric power systems using Monte Carlo methods. 24–30
62. Carvalho L, Issicaba D, da Rosa MA, Ramos JPV, Miranda V (2012) Reliability evaluation of generation systems via sequential population-based Monte Carlo simulation. IEEE 12th international conference on probabilistic methods applied to power systems (PMAPS), Istanbul, Turkey
63. Howard RA (1988) Decision analysis: practice and promise. Manag Sci 34:675–679
64. U.S. Department of Energy (2005) Wind power today. Federal Wind Program Highlights, Energy Efficiency and, Renewable Energy
65. Ensslin C, Milligan M, Holttinen H, O'Malley M, Keane A. Current methods to calculate capacity credit of wind power, IEA Collaboration. IEEE GM2008, Pittsburg, PA, USA
66. Garver LL (1966) Effective load carrying capability of generating units. IEEE Trans Power Appar Syst 85:910–919
67. Booth RR (1972) Power system simulation model based on probability analysis. IEEE Trans Power Appar Syst 91:62–69
68. D'Annunzio C, Santoso S. Analysis of a wind farm's capacity value using a non-iterative method. IEEE GM2008, Pittsburg, PA, USA
69. Huang SR, Chen SL (1993) Evaluation and improvement of variance reduction in Monte-Carlo production simulation. IEEE Trans Energy Conversion 8(4):610–619
70. Lim J, Jang J, Choi J, Cho K, Cha J (2012) Probabilistic production cost simulation and reliability evaluation of power system including renewable generators. iitmicrogrid.net
71. Miranda V, Carvalho LM, Rosa MA, Da Silva AML, Singh C (2009) Improving power system reliability calculation efficiency with EPSO variants. IEEE Trans Power Syst 24(4):1772–1779
72. Tatsuta F, Tsuji T, Emi N, Nishikata S (2006) Studies on wind turbine generator system using a shaft generator system. J Electr Eng Technol 1(2):177–184
73. White JA, Agee MH, Case KE (1989) Principles of engineering economic analysis. Wiley, New York

Chapter 5
Geographical Information Systems and Loop Flows in Power Systems

Manish Mohanpurkar, Hussein Valdiviezo Sogbi, and Siddharth Suryanarayanan

5.1 Introduction

ADVANTAGES of interconnected power systems include increased system security and reliability of supply, lowered operating costs, and increased Available Transfer Capacity (ATC). However, interconnections also introduce challenges, such as propagation of system events over wider areas and Unscheduled Flows (USFs) of electricity [1]. USFs represent the deviation of the actual power flowing on transmission lines from the market-scheduled flows. Contractual agreements between utilities are based on the fair market assumptions to optimize cost and operate electric grid at a desirable frequency. Deviation from market-schedules may also occur due to rerouting of power flows on account of inadvertent changes in the topology and may lead to forced participation of utilities and other assets that may not directly be involved in particular trades. USFs are known to reduce ATC, increase transmission losses with operation at or near stability limits, and complicate cleared transmission pricing [1]. Critical levels of USFs are mitigated by curtailment of schedules and deployment of qualified control devices in the system.

Alternatively non-critical levels of USFs can be accommodated in the market after clearance. One of the methods of accommodating USFs is based on linear estimation of *minor loop flows* [1]. This is done by employing a simple linear regression model to estimate the *minor loop flows* using the topology of the system, and the difference of the actual branch flows and the expected load flows [1]. A disadvantage of this method described in ref. [1], is that the topology matrix,

M. Mohanpurkar • S. Suryanarayanan (✉)
Department of Electrical and Computer Engineering, Colorado State University, Fort Collins, CO, USA
e-mail: suryanarayanan@gmail.com

H. Valdiviezo Sogbi
Universidad Autónoma de Yucatán, Mexico

E. Kyriakides et al. (eds.), *Electric Power Engineering Research and Education*, Power Electronics and Power Systems, DOI 10.1007/978-3-319-17190-6_5 also known as incidence or system matrix, that yields the relationship between the USFs and the *minor loop flows* is obtained visually. While this may be sufficient for smaller power grids and for proposing the accommodation method as a proof-of-concept, it may not be amenable to power grids of practical size. This chapter proposes an enhancement to the accommodation method proposed in ref. [1] by utilizing Geographical Information Systems (GIS).

Additionally, this chapter also shows the applicability of the accommodation method to scenarios of high penetration of a stochastic generation source, i.e., wind energy, in wide-area electric grids. This is considered nontrivial as the volatility in generation due to wind may manifest in USFs; and, in order to accommodate the USFs properly, there is a need for understanding this effect.

This chapter is organized as follows: Section 5.2 provides a brief note on GIS applications in power systems; Sect. 5.3 illustrates the prior work on the estimation of *minor loop flows* using USF measurements; Sect. 5.4 describes the application of the GIS techniques for synthesizing the system matrix for the above mentioned estimation problem; Sect. 5.5 applies this technique to a test system, and Sect. 5.6 describes the nature of variability in estimates as a function of wind penetration. Section 5.7 concludes the chapter.

5.2 Geographical Information Systems in Power Systems

GIS, which deals with the collection, management, and presentation of geographical data, is a widely used technique in infrastructure planning and management. Adapting GIS to power systems network planning and operational analysis possesses the potential of multiple advantages. GIS applications that provide visual representation of physical systems play a crucial role in determining the status and operational control of field devices. An application of GIS in the planning of a Medium Voltage (MV) distribution network of open loop configuration is discussed in ref. [2]. This is a prime example of a multi-objective optimization problem in distribution network planning and is solvable by various optimization methodologies. GIS have also found applications in operations and management of transmission and distribution networks due to the spatial nature of infrastructure as referred in the following section. The spatial and temporal functions are combined to detect distribution network faults in ref. [3]. Two classifiers, i.e., linear discriminant analysis and logistic regression, are then trained for evaluating the detected fault conditions to provide real time information. Essential steps to transform spatiotemporal functions of the network and time series data management to form a Device Data Management System (DDMS) are discussed in ref. [4]. A feature of integrating the visualization of realtime and historical events in the power system enables a more efficient operation; and, this employs GIS. Another optimization of operational cost for MV distribution networks is proposed in ref. [5], that functions by using smart meters serving as data inputs, and Automated Metering Infrastructure (AMI) serving as the communications media, whereas the GIS provides the interaction platform. For transmission networks, an online monitoring system is proposed using similar building blocks such as spatial functions, GIS information, and communication networks to construct intelligent power grids in ref. [6]. Management of congestion in bulk interconnections by mapping lines using GIS applications and congregating the various cost functions involved in the transmission of power is presented in ref. [7]. Congestion patterns, hourly nodal pricing, planning of transmission lines, and efficient integration of renewable energy sources are also discussed on the same platform. An exclusive interaction of system operators with the physical infrastructure via interfaces is developed based on GIS for efficient management of network infrastructure [8]. Thus, several applications in the implementation of Energy Management Systems (EMS) in power system networks using GIS exist.

5.3 USF Accommodation by Loop Flows

As mentioned in Sect. 5.1, a linear estimator is used to estimate *minor loop flows* for accommodating the USFs. The linear estimation equation is given as,

$$[H]_{nxp}[x]_{px1} = [z]_{nx1} \tag{5.1}$$

where H is the system matrix, z is the unscheduled flow on branches, x is the mathematical artifact, i.e., *minor loop flow* values, n is the number of branches in the network, and p is the total number of loop flows to be estimated. Eq. (5.1) is solved using the Ordinary Least Squares (OLS) technique and the pseudo-inverse of the non-square system matrix, H, to obtain the estimates of the loop flow, $\widehat{x}_{\text{ols}}$ as shown in Eq. (5.2).

$$[\widehat{x}_{\text{ols}}]_{px1} = [H^T H]_{pxp}^{-1}[H^T]_{pxn}[z]_{nx1} \tag{5.2}$$

The system matrix, H, may be considered as a combined representation of the closed minor loops of the network and the bidirectional transmission lines (branches or edges) of the network. The following are the rules assumed in ref. [1] while choosing the loops visually in order to maintain consistency in the selection process:

1. Each transmission line (branch) is assumed to be a bidirectional edge such that it can be traced in both directions.
2. Nodes with degree 1, i.e., nodes with only one connection/branch, especially generation nodes or sparsely located load points, cannot be a part of any loops.
3. Loops are chosen such that all lines other than single connection lines (see point no. 2) to be traversed twice, in opposite directions.
4. Loops with lower number of nodes are preferred over those with larger number of nodes, for the sake of simplicity.
5. There is no fixed sequence of choosing a starting node to form a loop, and hence the resulting loop may vary from analyst to analyst.
6. Loops are chosen solely based on the visual interpretation of the network.

The selection of loops may also be influenced by the manner in which the one-line diagram is drawn, making the accuracy and clarity of the one-line diagram a crucial aspect in the process. However, visual synthesis suffers from three basic drawbacks:

(a) Incorrectness in detecting loops for multi-planar graphs.
(b) Lack of accuracy checks.
(c) Inapplicability for topographically complex connections, i.e., practical bulk interconnections. For test systems of small to medium sized networks this technique is useful to especially analyze planning and operational scenarios associated with USFs.

Consider the following simple test system consisting of nine buses, shown in Fig. 5.1 [9]. Visual synthesis is sufficient enough to detect the minor loops in this network on account of its simplicity and small size, and is shown in Table 5.1.

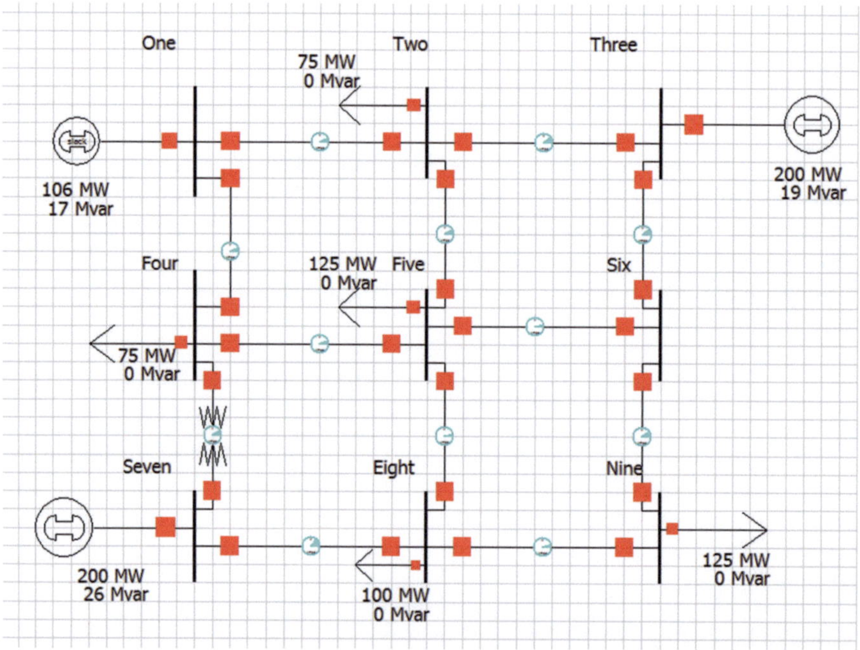

Fig. 5.1 A notional 9-bus test system directly adopted from [9]

Table 5.1 Visual synthesis of the loops of the notional 9-bus test system

Loop number	Nodes of the system in the loop
Loop 1	[1 2 5 4 1]
Loop 2	[3 6 5 2 3]
Loop 3	[4 5 8 7 4]
Loop 4	[6 9 8 5 6]

Note that all the non-peripheral transmission lines are scanned exactly twice, going in opposite directions. For example, the transmission line 2–5 is traced as 2 → 5 in *Loop 1* and as 5 → 2 in *Loop 2* as shown in Table 5.1. Using the loop information in the Table 5.1, the system matrix has to be synthesized; and, multiple methods can be employed to do so. In the first instance, the numerical inequality between the nodes of a branch can be explored. For example, for the branch 2–5 in the loop 1 scanned as 2 → 5, the entry "+*1*" may be assigned whereas for the same branch scanned a 5 → 2, an entry "−*1*" may be assigned for the apt element of the system matrix *H*. The logical operator approach uses the outcome of either "<" or ">" to obtain a pseudo-measure of allocating elements of system matrix. An alternative logic can be visual inferences of the location of transmission line layout with respect to reference directions. Visuals in Fig. 5.2 clarify this element assignment convention.

This logic draws from the visual interpretation of the network layout and can be regarded as an extension of the visual synthesis of loops. The system matrix entries using the visual synthesis for *Loop 1* (from Table 5.1) is given in Table 5.2. Table 5.3 provides the complete system matrix for the notional 9-bus test system using the discussed convention.

An enhancement over the above discussed procedure can be made using each edge exactly twice in opposite directions as a criterion for loop selection. The peripheral edges 1–2, 2–3, 3–6, 6–9, 9–8, 8–7, 7–4, and 4–1 are scanned only once to form loops. Another topographically valid loop can be formed by properly using all these unused edges. This loop represents the outermost periphery of the network and is not a minor loop. The loop selection problem is limited to a finite solution by exhausting the direction of scan of an edge when used once.

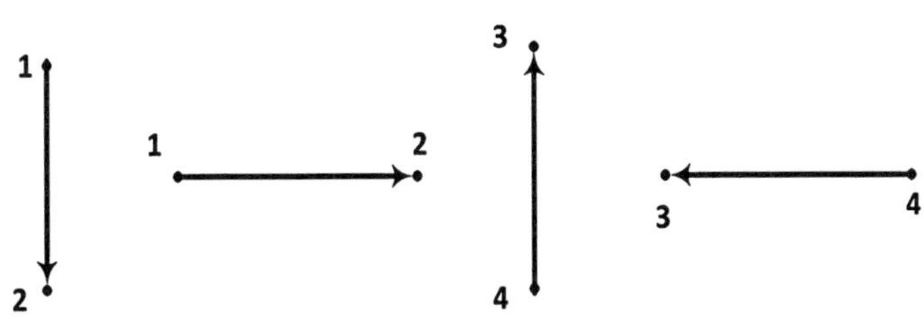

Fig. 5.2 Scan directions along the reference directions 1–2 will be allocated a "+*1*" and along the reference directions 3–4 "−*1*" will be allocated

Table 5.2 Visual synthesis of the system matrix elements corresponding to the *Loop 1*

Branches	1–2	2–5	5–4	4–1
Elements	+1	+1	−1	−1

Table 5.3 System matrix for the notional 9-Bus Test System using the convention shown in Fig. 5.1 [1]

Branches	Loop 1	Loop 2	Loop 3	Loop 4
1–2	1	0	0	0
1–4	−1	0	0	0
2–3	0	1	0	0
2–5	1	−1	0	0
3–6	0	1	0	0
4–5	−1	0	1	0
4–7	0	0	−1	0
5–6	0	−1	0	1
5–8	0	0	1	1
6–9	0	0	0	1
7–8	0	0	−1	0
8–9	0	0	0	−1

5.4 GIS Application in System Matrix Synthesis

The visual synthesis described in Sect. 5.3 requires human decision-making and intervention in both the selection of appropriate loops and the consequent synthesis of the system matrix. Visual synthesis is not suitable for bulk interconnections due to complexity and large size; hence an automated mechanism is needed. In order to replace the latter with an automated process, a new technique of using GIS coordinates of the physical location of the buses (nodes) is proposed. For the GIS approach, the layout information of the lines is obtained by processing of the coordinates of the respective nodes for the branch under consideration. Other related information that can be synthesized using the GIS coordinates includes: elevation, distance, and angles with respect to a reference axis. Distances between buses can be assumed as approximately equal to the Euclidean line lengths. The layout information of the line is more significant in this regard to help synthesize the system matrix. A two-step procedure to automate the loop detection and system matrix synthesis suitable to bulk interconnections is proposed.

Step 1. An algorithmic detection of loops in a power systems network.
Step 2. Use the loops obtained from Step 1 to form the system matrix in an automated manner.

For the context of this chapter, discussions pertaining only to the synthesis of the system matrix, i.e., Step 2, are given. Here, we assume that the input needed for executing Step 2, i.e., the nodes included in each minor loop, is known *a priori*. Discussion on the synthesis of the minor loops by algorithmic techniques is out of the scope of this chapter; however, such methods are currently under various stages of research and dissemination.

The dimension of the system matrix is (*nxp*), where *n* is the total number of branches and *p* is the total number of minor loops or regressors. Graphically, with respect to network, the rows represent the lines (branches) of the network in a specific order, which may be obtained from any commercial software database used for creating the power system case simulation. The most common way of representing the line data is by using numerical values of the buses (nodes) connected in an ascending order. However, exceptions such as displaying or storing power flow only in the positive directions (as indicated by Fig. 5.2), will lead to system-specific shifts in reordering branches. The columns of the system matrix correspond to the sequentially stored list of minor loops associated with the network (obtained from Step 1). The matrix elements corresponding only to the branches included in a particular loop will have a non-zero entry, whereas a branch not associated with a loop will have a zero entry [1].

Convention #1: The visual synthesis of loops can also provide the directional layout of the line and hence can be used along with a proper convention to synthesize the system matrix. A convention for a line layout being *either from the south to the north or from the west to the east* can be assumed to be a "+*1*" or "−*1*", whereas in opposite directions as "-*1*" or "+*1*" as applicable. This is a convenient method for small sized systems with accurate one-line diagrams and known directional information.

Convention #2: The intent of the method proposed in this chapter is to replace the human decision-making of determining the line layout by using the GIS coordinates. GIS coordinates are an inseparable component of standard bulk interconnection databases and are included from the planning stages. Using the GIS coordinates to obtain the *azimuth* to interpret the line layout information is proposed. The convention adopted here is based on the working database of respective networks (obtained either from standard dataset or commercial software packages) as: *from bus—to bus*. For this analysis the deviation of an actual transmission line from a straight line is ignored, i.e., the line length of a transmission line is assumed to be the same as Euclidean distance between the nodes. Transmission lines deviate from a straight line due to multiple reasons such as available right of way, topography of land, etc.

An *azimuth* is the angle, taken clockwise from north with a line between any two points. The North Pole has an *azimuth* of 0° from every other point on the globe [8]. The interpretation of an *azimuth* is provided by a simplified example below:

The GIS coordinates of point 1 are (latitude; longitude) = (21° 0′0″; 89° 30′0″) and the point 2 = (40° 30′0″; 105° 0′0″). The *azimuth* of 319° is obtained for the straight line obtained by joining point 1 to point 2. The distance between the two points is 1,632.342 miles. The distance so obtained might not be accurately equal to the line length as given in the database since deviation of actual lines from straight line is frequent. However, the intended purpose of setting up the system matrix does not require the distances between the buses at all, and hence the inconsistency between the geographical distance and line length can be ignored. The North Pole is used as the reference direction to obtain directional information. The loops obtained from the visual synthesis and the directional information are the two components used to synthesize the system matrix. Figure 5.3 shows the convention used to allocate the elements of the system matrix.

In the loops chosen, for any branch of the network either of the nodes can be a reference point with the other point being the endpoint. The straight-line layout of this line segment with respect to the North Pole is obtained and the value of "+*1*" or "−*1*" is inserted at the appropriate element in the system matrix. A uniform convention needs to be adopted if the line segment lies exactly on a boundary of

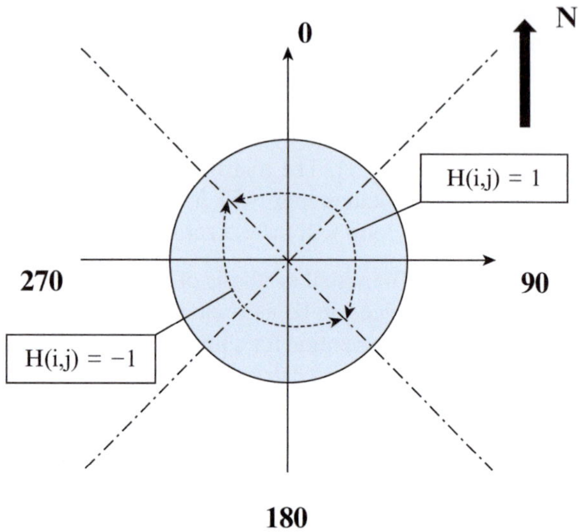

Fig. 5.3 Convention for synthesizing the system matrix elements using the *azimuth* (in degrees) of the edges in loops.

Table 5.4 Convention adopted for assigning system matrix elements using GIS coordinates and directional information

Azimuth of an edge (x°)	Value
$0^{\circ} < x^{\circ} \leq 135^{\circ}$	1
$135^{\circ} < x^{\circ} \leq 315^{\circ}$	−1
$315^{\circ} < x^{\circ} \leq 360^{\circ}$	1

the two regions i.e., coincides with the dashed lines shown in Fig. 5.3. Table 5.4 shows that if the line segment has an angle of 135° then a "+*1*" will be allocated, which implies that the direction of the segment is eastward which is positive. The exact reverse convention would be used for allocating a "−*1*" such that the segment layout is assumed to be southward. Similar argument can be made for a line segment with an angle of 315°, which will be allocated an element of "−*1*" assuming that the direction of this segment is westward. Depending on the magnitude of this angle, we are going to build the system matrix and use "+*1*" or "−*1*" according to Table 5.4, where x° is the *azimuth* of the line.

All the loops of a network are processed such that each row of the system matrix will have only two non-zero entries i.e., "+*1*" and "−*1*". Generation and load buses with degree *1* will have all-zero entries. Eventually, these rows will be removed from the system matrix as they do not provide any notable information about loop flows.

5.5 Demonstration and Discussions

The IEEE 14-bus test system is used to demonstrate the result of synthesizing the directional information of the branches in distinct loops . Using the angle (*azimuth*) of the transmission lines (branches) and the sequence of scan of the branches, a suitable system matrix is synthesized in Table 5.5 with the intermediate results displayed in Tables 5.6, 5.7, and 5.8. A set of loops is obtained by visual synthesis and using the rules of choice as explained in Sect. 5.3. The test system comprises of 14 nodes, 20 branches, 6 generators, and 11 loads and its graph equivalent is shown in Fig. 5.4 [10]. Appendix provides the details of the generator, load, and line flow values corresponding to the base case for the test system. The market expected line flows are directly assumed to be the base case line flows since they represent the most likely scenario. Visually selected loops are as shown in Table 5.8. The standard database of the IEEE 14-bus test system does not specify the line lengths; hence, they are to be chosen rationally. Bus no. 1 is proposed as a point of reference to determine the latitudes and longitudes of the other buses (nodes) using assumed line lengths as a factor. In this example, bus no. 1 of the test system is assumed to be located in Fort Collins, CO (40.60° N, 105.13° W). With these coordinates, and the assumed lengths of each branch and test system knowledge, we can calculate the latitude and longitude for all the buses (nodes). For this, we propose a *rhumb* direction for each one of the nodes, and with this direction and the distance we get the coordinates. A *rhumb* line crosses all the meridians of longitude at the same angle. The approximate line lengths assumed and the corresponding *rhumb* line angles are shown in Table 5.6. Table 5.7 provides the network information along with the coordinate locations (derived from the assumed line lengths). Table 5.8 depicts the inclination, i.e., *azimuth* (in degrees) of each edge in loops with respect

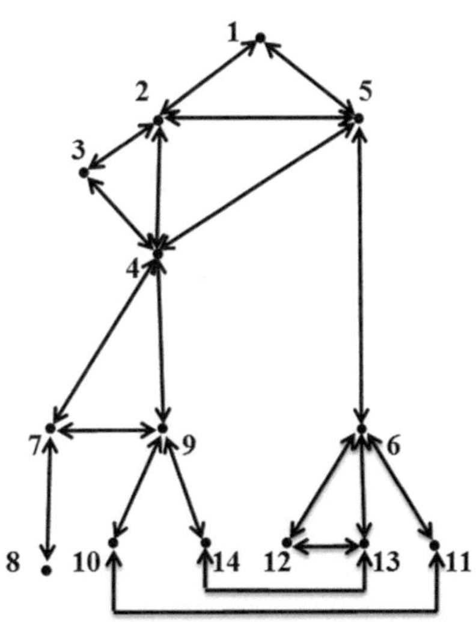

Fig. 5.4 Bidirectional graph equivalent of the IEEE 14 bus test system

Table 5.5 System matrix for the IEEE 14 Bus Test System using the GIS coordinates

	Minor loop #									Minor loop #							
Branch	1	2	3	4	5	6	7	8	Branch	1	2	3	4	5	6	7	8
1	1	0	0	0	0	−1	0	0	11	0	0	0	0	1	0	0	−1
2	1	0	0	0	0	−1	0	0	12	0	0	−1	0	0	1	0	0
3	0	−1	0	1	0	0	0	0	13	0	0	1	0	0	0	0	−1
4	0	1	0	0	0	−1	0	0	14	0	0	0	0	0	0	0	0
5	−1	0	0	1	0	0	0	0	15	0	0	0	0	0	1	−1	0
6	0	1	0	−1	0	0	0	0	16	0	0	0	0	1	0	0	−1
7	0	0	0	−1	1	0	0	0	17	0	0	0	0	0	1	0	−1
8	0	0	0	0	0	−1	1	0	18	0	0	0	0	−1	0	0	1
9	0	0	0	0	1	0	−1	0	19	0	0	1	0	0	−1	0	0
10	0	0	0	0	−1	1	0	0	20	0	0	0	0	0	1	0	−1

Table 5.6 Assumed line lengths in miles and *Rhumb* line angle values

Node 1	Node 2	Line lengths (miles)	*Rhumb* line angle (in degrees)	Node 1	Node 2	Line length (miles)	*Rhumb line angle* (in degrees)
1	2	120	200	6	11	50	105
1	5	100	110	6	12	50	250
2	3	80	230	6	13	50	195
2	4	100	160	7	8	110	255
2	5	157	59.22	7	9	64	102.68
3	4	105	113.64	9	10	50	185
4	5	170	113.64	9	14	60	130
4	7	100	230	10	11	240	78.4
4	9	80	190	12	13	46	132.21
5	6	250	165	14	13	130	89.33

Table 5.7 Proposed coordinates of the IEEE 14-bus test system (system assumed to be located in the U.S.)

Node	Characteristic components	Coordinates	Node	Characteristic components	Coordinates
1	Reference and slack generator	40.6° N, 105.13° W	8	–	36.22° N, 108.54° W
2	Conventional generator and load	38.96° N, 105.8° W	9	Load bus	36.44° N, 105.51° W
3	Load bus	38.21° N, 107.01° W	10	Load bus	35.75° N, 105.58° W
4	Load bus	37.59° N, 105.26° W	11	Load bus	36.38° N, 101.31° W
5	Wind farm and load bus	40.09° N, 103.35° W	12	Load bus	36.32° N, 103.02° W
6	Wind farm and load bus	36.58 °N, 102.18° W	13	Load bus	35.87° N, 102.41° W
7	Wind farm	36.65° N, 106.64° W	14	Load bus	35.87° N, 104.68° W

to the North Pole measured in the clockwise direction, between the starting node and the ending node of the edge. These entries can have a value from 0° to 360°. The first row of Table 5.8 shows the *azimuth* values for the three edges in *Loop 1*, i.e., 2–1, 1–5, and 5–2. Sequential index is the total edges associated with respective loops. The largest is the sixth loop with ten edges. Using the convention explained in Fig. 5.3 and Table 5.8, the system matrix for the IEEE 14-bus test system is obtained as shown in Table 5.5. The system matrix has 38 non-zero elements. The total branches considered are 20; and, since the branch 7–8 (branch 14) is not traversed in any loop we will discard it. The remaining 19 branches have been traced twice in opposite directions, and hence we have the 38 non-zero elements. The system matrix has a sparsity index of 25 %. As mentioned earlier, in standard

Table 5.8 Visually selected loops and *azimuth* values for edges associated with the loops listed sequentially in the IEEE 14-bus test system

		Sequential edge index									
Loop number	Nodes of the system in the loop	1	2	3	4	5	6	7	8	9	10
Loop 1	[2 1 5 2]	19.36	110.02	240.84	0	0	0	0	0	0	0
Loop 2	[2 3 4 2]	229.76	113.64	340.34	0	0	0	0	0	0	0
Loop 3	[6 12 13 6]	249.20	132.21	14.58	0	0	0	0	0	0	0
Loop 4	[4 3 2 5 4]	294.72	49.07	59.22	211.36	0	0	0	0	0	0
Loop 5	[4 5 6 11 10 9 4]	30.16	164.98	105.70	260.92	4.67	9.77	0	0	0	0
Loop 6	[4 7 9 14 13 12 6 5 1 2 4]	229.91	102.69	130.14	89.33	312.57	68.70	345.71	291.18	199.85	159.95
Loop 7	[4 9 7 4]	189.92	283.36	49.08	0	0	0	0	0	0	0
Loop 8	[9 10 11 6 13 14 9]	184.71	78.40	286.21	194.71	270.67	310.63	0	0	0	0

power systems database the coordinates of buses are available and hence it is relatively easier to use this approach. In such cases, the *azimuth* can be directly calculated and along with the proposed convention as explained in Sect. 5.3, the system matrix can be synthesized automatically.

5.6 Estimated USFs and Variability

5.6.1 Experimental Setup

A Monte Carlo simulation is set up for the IEEE 14-bus test system in order to analyze the variability induced in the USFs as a result of penetration of wind energy sources. For this purpose, four wind farms of installed capacity approximately 222, 99, 95.9, and 49.8 MW are assumed to be connected one each at buses 5 and 7 and two at bus 6, respectively. The physical distances between the wind farms are approximately equal to the assumed line lengths. The wind farms are assumed to have local voltage regulation at respective buses. Real power injection by the wind farms measured at the point of interconnection for a year is used as the variable input to the power flow. All other parameters pertinent to the power flow algorithm are maintained constant to execute 500,000 iterations in the Monte Carlo simulation. The base case is drawn using the average values of wind farm penetration as a moderate scenario. The fixed load and conventional generation values used in the simulation are as tabulated in the Appendix. The slack generator is expected to supply the transmission losses as well as serve as the secondary market for wind farms to supply deficits or account for surfeits, if any. The vector of unscheduled power flows (z) is computed as the difference between the market expected flow (base case flows) and actual power flows obtained from the Monte Carlo simulation. OLS estimates are computed using the Pseudo-inverse technique as shown in Sect. 5.3.

5.6.2 Variability in USFs

Following are the histograms for the estimated *minor loop flows* listed in Table 5.8 under the annual heavy wind energy penetration scenario. Figs. 5.5, 5.6, 5.7, and 5.8 display the histograms of the estimated loop flows, for *Loops 1*, *5*, *7*, and *8*, respectively, as a representative sample. The plots indicate a pattern of having a concentrated probability near specific MW values for the chosen inputs i.e., an approximate unimodal distribution. The value of this USF corresponds to the most frequent penetration of the wind farm in the networks. Another observation about the wind farm power outputs and the USFs has no fixed cause and effect relation. USFs and wind farm penetration exhibit both positive and negative correlation. Table 5.9 shows the values of correlation found between the different estimated minor loop flows and the wind farm penetrations obtained from the varying input.

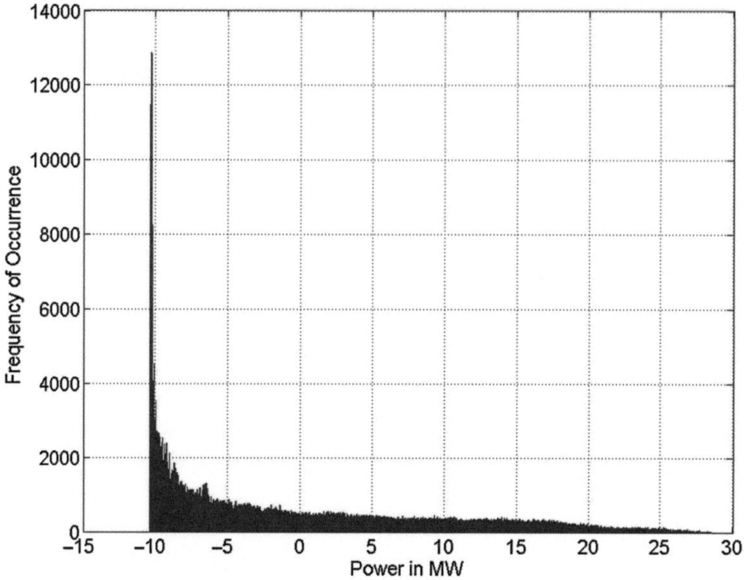

Fig. 5.5 Histogram of estimated loop flow for *Loop 1* using OLS

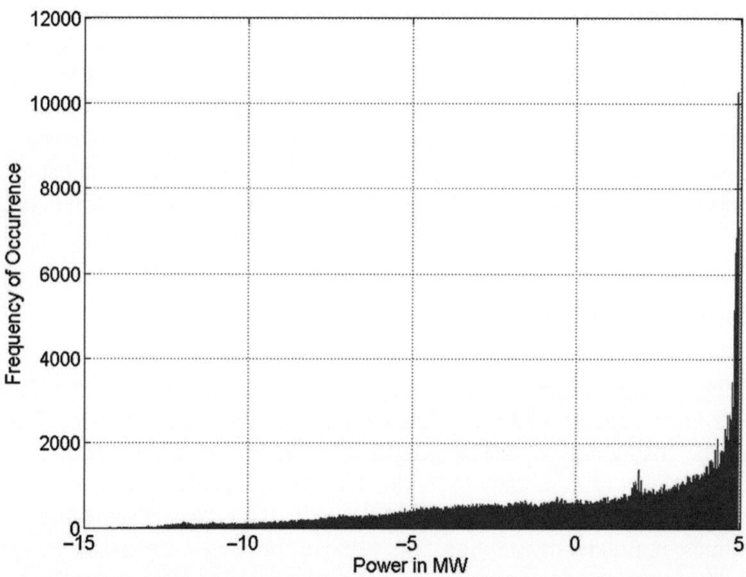

Fig. 5.6 Histogram of estimated loop flow for *Loop 5* using OLS

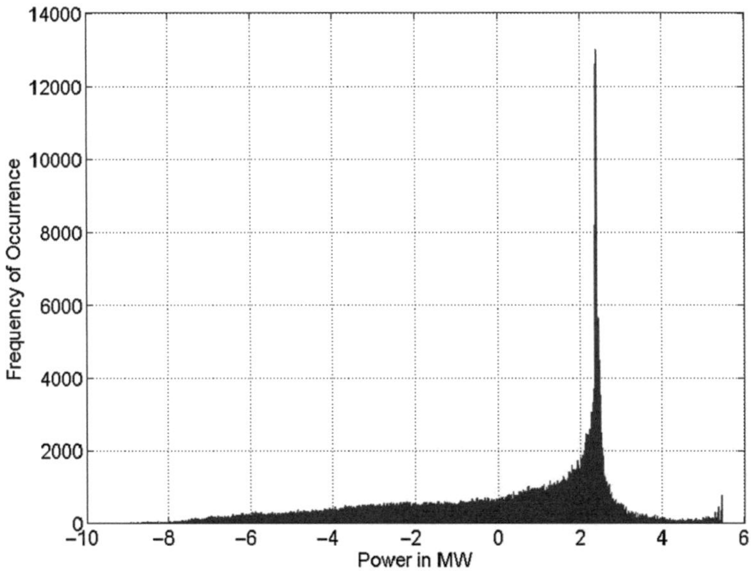

Fig. 5.7 Histogram of estimated loop flow for *Loop 7* using OLS

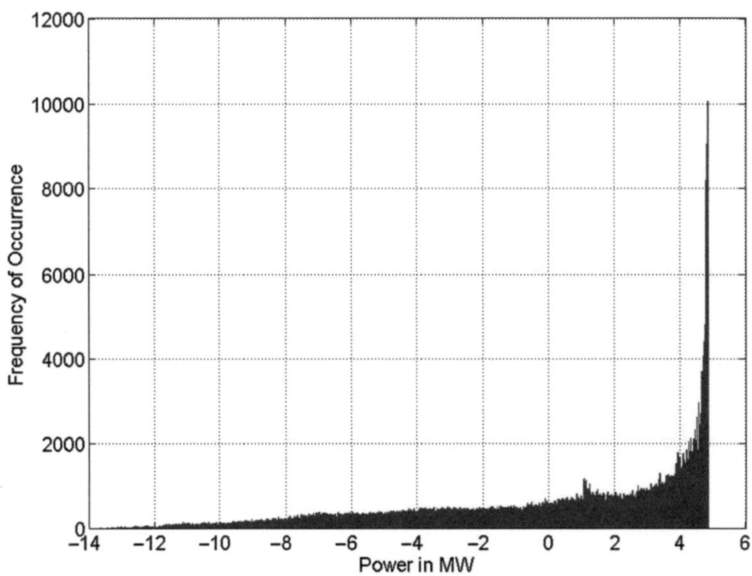

Fig. 5.8 Histogram of estimated loop flow for *Loop 8* using OLS

Table 5.9 Correlation between USFs and wind farm penetrations at different buses

Loop number	Wind farm output at bus 5	Wind farm output 1 at bus 6	Wind farm 2 output at bus 6	Wind farm output at bus 7	Variance of loop flows
Loop 1	0.92	0.79	0.68	0.59	101.55
Loop 2	0.8989	0.80	0.69	0.63	31.10
Loop 3	−0.936	−0.74	−0.64	−0.67	27.90
Loop 4	0.90	0.77	0.67	0.67	33.09
Loop 5	−0.71	−0.88	−0.74	−0.55	19.85
Loop 6	−0.96	−0.65	−0.57	−0.71	25.73
Loop 7	−0.90	−0.73	−0.66	−0.26	8.23
Loop 8	−0.88	−0.78	−0.68	−0.67	19.86

This is highly specific to the test system and the market-scheduled flow inputs chosen. No generic inferences can be drawn from this analysis as yet.

As seen from Table 5.9, the output of wind farm at bus 5 has a strong positive correlation with estimated *Loop 1* whereas a strong negative correlation with the estimated *Loop 3*. The value of the correlation index is a function of both the degree of connectivity of the bus and the scheduled-market flows on lines connected to that bus. The largest annual variance is observed for the estimated loop flow associated with *Loop 1*, and it may be attributed to the large variance of wind farm power output at bus 5. Additionally, the installed capacity of this wind farm is significantly larger than the others, thus imparting comparatively greater variability to the estimated loop flow. Similar inferences can be drawn for the rest of the estimated loop flows and wind farm outputs after further investigation. The location of the wind farms is crucial in this study such that the inferences drawn from the estimated values will change significantly, if they are connected to different buses. It is also noted, in practical cases the vector of unscheduled flows (z) should be computed after every market closure as both actual power flow and market expectations are bound to change. These may be computed on an hourly or to a few minutes resolution depending on the market type. For the sake of simplicity, such minute details are not accounted but are certainly necessary. This concludes the demonstration of the application of GIS based synthesis of system matrix to accommodate USFs using *minor loop flows*.

5.7 Concluding Remarks

An application of a GIS technique in estimating *minor loop flows* in wide area power grids is explored. A simple linear estimator is used to estimate the *minor loop flows* using the system information and measurements of Unscheduled Flows. Line layouts within loops are synthesized using the GIS coordinates of the buses to automate the formation of the system (incidence) matrix. Unscheduled Flows on transmission lines (branches) are accommodated using the estimates of the loop flows, which in this case are estimated using Ordinary Least Squares. The system matrix for the IEEE 14-bus test system was synthesized for estimating loop flows in an annual variable generation scenario. Loop flow estimates show positive and negative correlations with wind farm output depending upon the remoteness of loops from the wind farm bus, but no fixed pattern was discernible.

Acknowledgment This work was funded in part by the Western Electricity Coordinating Council (WECC) as a subcontract under contract DOE-FOA0000068, and in part by the CONACYT Mexico Scholarship.

The authors thank Dr. G.T. Heydt, Regents' Professor in the School of Electrical, Computer, and Energy Engineering at Arizona State University, for his input to our study of loop flows. It is noteworthy that Profs. Heydt and Suryanarayanan developed the early framework for the accommodation of USFs in collaboration with Prof. R.G. Farmer (deceased, of ASU) and Mr. S. Chakka (formerly of ASU) [1], [11–15]. Dr. Mohanpurkar and Prof. Suryanarayanan further developed a robust statistical framework to estimate loop flows in electric grids with high wind penetration [16–18]. We also thank Mr. D.J. Zimmerle, of Colorado State University, for his inputs on graph theory and loop selection techniques. Wind data from the National Renewable Energy Laboratory, Golden, Colorado is acknowledged.

Disclaimer This chapter was originally published in the proceedings of the *45th Frontiers of Power Conference* and presented at the same conference in 2012 at Stillwater, OK. The authors have received full copyright waiver from the organizers of the above conference to publish this material.

Appendix

The IEEE 14 bus test system has been directly taken from [10] with the addition of wind farms.

(a) Bus-wise load information

Bus number	Load MW	Load MVAr
1	–	–
2	21.74	12.7
3	94.2	19
4	47.8	−3.7
5	7.6	1.6
6	11.2	7.5
7	–	–
8	–	–
9	29.5	16.6
10	9	5.8
11	3.5	1.8
12	6.1	1.6
13	13.5	5.8
14	14.9	5

(b) Average generation values

Bus number	Gen Id	Gen MW	Gen MVAr
1	1	186.9	3.82
2	1	40.04	87.45
5	1	20.73	−4.95
6	1	10.65	−13.86
6	2	7.38	−13.86
7	1	4.48	−49.45

(c) Base case line flows

From bus	To bus	Power flow MW	MVAr
1	2	129.8	130.4
1	5	57.1	59.4
2	3	68.1	72.4
2	4	47.2	49.5
2	5	29.9	34.3
3	4	−28.4	28.4
4	5	−70	70.7
4	7	23.9	25.4
4	9	15.2	15.5
5	6	27.2	40.7
6	11	8.4	12.9
6	12	7.6	7.6
6	13	18.1	18.1
7	8	0	0
7	9	28.3	50.7
9	10	4.5	18.8
9	14	9.7	15.9
10	11	−4.6	13
12	13	1.4	1.7
14	13	−5.6	9

References

1. Suryanarayanan S, Farmer RG, Heydt GT, Chakka S (2004) Estimation of unscheduled flows and contribution factors based on Lp norms. IEEE Trans Power Syst 19(2):1245–1246
2. Kong T, Cheng H, Hu Z, Yao L (2009) Multiobjective planning of openloop MV distribution networks using ComGIS network analysis and MOGA. Electr Power Syst Res 79(2):390–398
3. Cai Y, Chow MY (2009) Exploratory analysis of massive data for distribution fault diagnosis in smart grids. In: 2009 Proceedings of IEEE PES General Meeting, pp 1–6

4. Parikh PA, Nielson TD (2009) Transforming traditional geographic information system to support smart distribution systems. In: 2009 Proceedings of IEEE PES Power Systems Conference and Exposition, pp 1–4
5. Mak ST, Farah N (2012) Synchronizing SCADA and Smart Meters operation for advanced smart distribution grid applications. In: 2012 I.E. PES Innovative Smart Grid Technologies (ISGT), pp 1–7
6. Wu K, Zhang Z (2010) Research and implementation of smart transmission grids based on WebGIS. In: 2010 Proceedings of Second International Conference on Communication Systems, Networks, and Applications (ICCSNA), vol. 1, pp 302–307
7. Albaijat M, Aflaki K, Mukherjee B (2012) Congestion management in WECC grid. In: 2012 I. E. PES Innovative Smart Grid Technologies (ISGT), pp 1–8
8. Bonham-Carter GF (1994) Geographical information systems for geoscientist: modeling with GIS. Pergamon, Ottawa, ON
9. Chakka S, Suryanarayanan S, Heydt GT (2002) Analysis and estimation of loop (parallel) flows in wide area interconnected power systems. In: 2002 Proceedings of 34th North American Power Symposium (NAPS), pp 176–180
10. Allan RN, Al-Shakarchi MR (1977) Probabilistic techniques in a.c. load-flow analysis. Proc Inst Electr Eng 124(2):154–160
11. Suryanarayanan S (2008) Techniques for managing unscheduled flows in electric power systems and markets. In: 2008 Proceedings of IEEE Power and Energy Society General Meeting, Pittsburgh, PA, pp 1–6
12. Suryanarayanan S, Montgomery DC, Heydt GT (2005) Considerations for implementing tag schedules in transmission circuits. IEEE Trans Power Syst 20(1):523–524
13. Suryanarayanan S, Heydt GT, Farmer RG, Chakka S (2004) An estimation technique to assign contribution factors for loop flows in an interconnected power system. Electr Power Compo Syst 32(8):813–826
14. Suryanarayanan S, Heydt GT (2008) Modification to contribution factor formula for unscheduled flows. IEEE Trans Power Syst 23(2):809–810
15. Suryanarayanan S (2004) Accommodation of loop flows in competitive electric power systems, PhD dissertation, Department of Electrical Engineering, Arizona State University (advisor: Heydt GT)
16. Mohanpurkar M (2013) Computation of loop flows in electric grids with high wind energy penetration. PhD dissertation, Department of electrical and computer engineering, Colorado State University (advisor: Suryanarayanan S)
17. Mohanpurkar M, Suryanarayanan S (2013) Accommodating unscheduled flows in electric grids using the analytical ridge regression. IEEE Trans Power Syst 28(3):3507–3508
18. Mohanpurkar M, Suryanarayanan S (2014) Regression modeling for accommodating unscheduled flows in electric grids. IEEE Trans Power Syst 29(5): 2569–2570

Chapter 6
Introduction to Transmission Expansion Planning in Power Systems

Hui Zhang

Nomenclature

a_g	Quadratic cost coefficient of generator g
b_g	Linear cost coefficient of generator g
b_k	Series admittance of line k, a negative value
b_{k0}	Charging admittance of line k
c_g	Fixed cost coefficient of generator g
c_k	Investment cost of the line k
CF_{gt}	Capacity factor of generator g in year t
CG_{gt}	Hourly energy cost of generator g in year t
d	Discount factor
g_k	Conductance of line k, a positive value
$k(l)$	The slope of the lth piecewise linear block
M	Disjunctive factor, a large positive number
ng	Total number of generators
nl	Total number of lines, including potential lines
P_k	Active power flow on line k
$\Delta P_k(l)$	The lth linear block of active power flow on line k
PD_d	Active power demand of load d
PG_g	Active power generated by generator g
$\mathrm{PG}_g^{\max}$	Maximum active power output of generator g
$\mathrm{PG}_g^{\min}$	Minimum active power output of generator g
PL_k	Active power loss on line k
Q_k	Reactive power flow on line k

H. Zhang (✉)
California ISO, Folsom, CA 95630, USA
e-mail: hzhang@caiso.com

E. Kyriakides et al. (eds.), *Electric Power Engineering Research and Education*,
Power Electronics and Power Systems, DOI 10.1007/978-3-319-17190-6_6

QD_d	Reactive power demand of load d
QG_g	Reactive power generated by generator g
QG_g^{max}	Maximum reactive power output of generator g
QG_g^{min}	Minimum reactive power output of generator g
QL_k	Reactive power loss on line k
r_k	Series resistance of line k
S_k^{max}	MVA rating of line k
SD_d	MVA of load d
TO	Operating horizon
TP	Planning horizon
V_i	Bus voltage magnitude in p.u. at bus i
ΔV_i	Voltage magnitude deviation from 1.0 p.u. at bus i
ΔV^{max}	Upper bound on the voltage magnitude deviation
ΔV^{min}	Lower bound on the voltage magnitude deviation
x_k	Series reactance of line k
y_k	Series admittance of line k
y_{k0}	Charging admittance of line k
z_k	Binary decision variable for a prospective line k
Z_k	Series impedance of line k
$u_k(l)$	Binary variable for the lth linear block
δ_k	Binary variable for modeling $\|\theta_k\|$
θ_k	Phase angle difference across line k
θ^{max}	Maximum angle difference across a line
θ_k^+, θ_k^-	Nonnegative slack variables used to replace θ_k
$\Delta\theta_k(l)$	The lth linear block of angle difference across line k
Ω_g	Set of generators
Ω_k	Set of existing lines
Ω_k^+	Set of prospective lines

6.1 Background

The national push for a smart grid and the increasing penetration of renewable energy resources today has significantly influenced the operations and planning of the traditional power system. The future power grid is expected to be a smarter network that is flexible and robust enough to withstand various uncertainties and disturbances. According to the 10-year planning summary prepared by the Western Electricity Coordinating Council (WECC), loads are projected to increase 14 % from 2009 to 2020, which is a 1.2 % compound annual growth rate [1]. From the generation side, the future generation mix is expected to have a significant departure from the past because the addition of new generation to replace the retired units is dominated by renewables to fulfill state-mandated renewable portfolio standards (RPSs). By the year 2020, a total amount of 15 GW in nameplate generation is going to retire and 59 GW of additional generation will be added in the US Western Interconnection. Among the cited 59 GW, over 50 % is composed of wind and solar

PV. In addition, the US Western Interconnection is projected to generate 17 % of its energy from non-hydro renewable sources in 2020.

With these contemporary changes, some problems are expected in the future power system. First, the load increase may change the power flow in the existing grid and may result in potential overloads and stability issues. These issues may violate reliability criteria. Second, the renewable resources are usually located in remote areas and are not readily connected to the main power grid. In order to address these problems, additional transmission capacity is needed.

TEP is an important research area in power systems and has been studied extensively during the past several decades. The TEP exercise normally focused on improving the reliability and security of the power system when economic impacts were not the primary concern. In contemporary power systems however, the increasing complexity of the network structure and the deregulated market environment have made the TEP problem a complicated decision-making process that requires comprehensive analysis to determine the time, location, and number of transmission facilities that are needed in the future power grid. Building the correct set of transmission lines will not only relieve congestions in the existing network, but will also enhance the overall system reliability and market efficiency. The state of the art of the TEP studies is reviewed and summarized below:

Various TEP models have been developed during the past several decades. Among these models, mathematical programming and heuristic methods are two major classes of solution approaches. Mathematical programming methods guarantee the optimality of the solution in most cases, but often have stricter requirements on the model to be optimized. In order to obtain the global optimal solution efficiently, the problem or at least the continuous relaxation of the problem should have a convex formulation. Heuristic methods, on the other hand, are usually not sensitive to the model to be optimized and can potentially examine a large number of candidate solutions. The main criticism of heuristic methods, however, is that most of such methods do not guarantee an optimal solution, and provide few clues regarding the quality of the solution. Reference [2] presents a comprehensive review and classification of the available TEP models.

Due to the complexity of the TEP problem, the DC power flow model has been extensively used for developing TEP models [3–9]. One of the early works, [3], presents a linear programming (LP) approach to solve TEP problems. A mixed integer linear programming (MILP) based disjunctive model in [4] eliminates the nonlinearity caused by the binary decision variables. In [6], the behavior of the demand was modeled through demand side bidding. A bilevel programming model appears in [7] where the solution to the problem is the Stackelberg equilibrium between two players. A transmission switching coordinated expansion planning model was presented in [8] where the planning problem and the transmission switching problem are solved alternately. In terms of security constraints, the North American Electric Reliability Corporation (NERC) planning criteria state that power systems must survive an $N-1$ contingency [10]. For the linearized model, this criterion simply indicates that there should be no thermal limit violation with the outage of a single transmission or generation facility. The modeling of security constraints can be found in [11–13], where an MILP based disjunctive method is proposed for transmission line switching studies.

The active power losses are usually neglected in the linearized power flow model. However, the losses may shift the generation economic dispatch solution and therefore influence the optimal transmission plan. Two loss models are presented in [5] and [11], where the proposed models use piecewise linear approximations to represent the quadratic loss term.

Application of the AC power flow model to TEP problems (ACTEP) is rarely discussed in the literature. The advantage of formulating TEP problems using the AC model is that the AC model represents the electric power network accurately. Nevertheless, the nonlinear and non-convex nature of the ACTEP model can make the problem very difficult to solve and to obtain a desirable solution. Reference [14] presented a mixed-integer nonlinear programming (MINLP) approach for solving TEP problems using the AC network model. The interior point method and a constructive heuristic algorithm were employed to solve the relaxed nonlinear programming problem and obtain a good solution. It is reported in [15] that by relaxing binary variables, the NLP-based ACTEP model can solve a small-scale TEP problem within an acceptable time range and obtain a local optimal solution. However, solving a MINLP-based ACTEP problem is still extremely challenging at this moment.

Heuristic approaches are an alternative to mathematical programming for solving optimization problems. Heuristic approaches usually refer to the algorithms that mimic some behavior found in nature, e.g., the principle of evolution through selection and mutation (genetic algorithms). For problems that have significant computational complexity in finding an optimal solution, heuristic methods can usually give a solution with relatively smaller computational effort, though the obtained solution may not be optimal. In recent years, heuristic methods have been introduced to solve TEP problems in power systems [16–21]. In many of these instances, the heuristic method is not used on a stand-alone basis. In order to obtain better computational performance, heuristic methods are frequently used in conjunction with mathematical methods when solving practical TEP problems.

Some renewable generation resources such as wind and solar PV can be highly unpredictable. These renewable generation sources, if massively integrated, could greatly affect the power system operations and undermine the grid reliability. The traditional TEP models are based on a deterministic framework where loads are treated as known fixed parameters. The deterministic model certainly simplifies the problem, but fails to capture the stochastic nature of the real power system and may generate unrealistic transmission plans. In recent years, modeling of uncertainties in the TEP model has drawn increasing attention [22–29]. Stochastic programming, chance-constrained programming and scenario-based analysis are three approaches that are frequently used.

Two-stage stochastic programming is a widely used stochastic formulation that optimizes the *mathematical expectation* of the weighed future scenarios. A two-stage stochastic programming-based TEP model is proposed in [22] to coordinate the generation and transmission planning. In [23], a scenario-based multi-objective TEP model is presented to address the uncertainties and risks in the planning process. Due to the computational burden, decomposition methods are usually used to solve the above stochastic TEP models [24]. In terms of the resource uncertainties, a probabilistic power flow (PPF)-based planning model is proposed in [25]. That reference evaluates a statistical range of the possible power flows instead of a single solution. A chance-constrained model is presented in [26] to address the uncertainties of loads and wind farms. It should be noted that the PPF-based planning model and the chance-constrained planning model are both risk-based games in which the planners need to decide the confidence level at a specified risk. In terms of reliability assessment, the probabilistic approach can also be applied [27, 28]. The traditional deterministic planning approaches are not able to capture the probabilistic characteristics in power system. In reality, this may lead to either overinvestment or potential reliability violations [27]. A method for choosing the optimal expansion plan considering a probabilistic reliability criterion is proposed in [28]. The probabilistic planning concept is applied to liberalized electricity markets in [29].

Compared to the static planning model where lines are planned for a single target year, the multi-stage planning model considers the continuing growth in demand and determines *when* to carry out the transmission expansion as well [30, 31]. The major obstacle in the development of multi-stage planning models is still the computational burden. Heuristic algorithms are usually used to solve multi-stage planning problems. A genetic algorithm is presented in [30] to solve the problem of multistage and coordinated TEP problem. A multi-criteria formulation for multiyear dynamic TEP problems is presented in [31] and is subsequently solved by a simulated annealing algorithm with the objective to find the optimal balance of investment costs, operation costs, as well as the expected unsupplied energy. Ordinal optimization is used in [32] for solving a multi-year TEP problem. The ordinal optimization algorithm uses crude models and rough estimates to derive a small set of optimal plans in each sub-planning period for which simulations are necessary and worthwhile to find acceptable solutions. In [33], a multiyear security constrained generation-transmission planning model is presented and a constructive heuristic algorithm is developed to solve the problem.

6.2 The TEP Framework

The goal of the contemporary TEP exercise is to improve the overall market efficiency and enhance system reliability. In order to balance the economics and the reliability, it is crucial to identify the set of transmission lines to be added. Beyond the traditional power flow and contingency analysis, the TEP exercise today is under transition to a comprehensive decision-making process in which the value of the transmission projects needs to be accurately evaluated.

The traditional TEP approach focused on protecting the system from the "worst case" scenario. In other words, the transmission network is designed to protect against the worst contingency at the peak load level. This approach was based on the assumption that if a system survived the worst contingency, the system would be robust enough to survive any other contingency. While this assumption might be valid in some cases, the worst case-based TEP approach is not suited for contemporary power systems. The reasons are twofold: first, some critical contingencies are very unlikely to occur; protecting against these contingencies by building more transmission lines is not economical. Instead, remedial action schemes (RASs) are usually designed to mitigate the impact of these contingencies by sequentially opening a set of lines or even tripping generation or loads. Second, the massive installation of renewable resources and the deepening of deregulated electricity markets have brought increasing uncertainties to the power grid. These factors make the "worst case" difficult to identify because it may not always occur at peak load level. In addition, the system operating costs are also subject to randomness. Prices of different resources such as gas and coal are unclear in the future. Inflation and possible delays in completion can affect the estimated benefits as well. All these factors make an exact transmission expansion planning process, without considering uncertainties, unrealistic. Thus, considering uncertainties in deciding transmission expansion investments is critical.

With the above factors considered, the proposed framework for next-generation TEP is shown in Fig. 6.1. The framework can be categorized into four stages as described in the dotted boxes. In order to develop a planning base case in the first stage, one can consider the operational case of the current year, adjust the load level according to the forecast of the load growth, remove the generators to be retired and add the generators to be installed in the targeted planning year. Federal policy requirements, e.g., RPS, and stakeholder's inputs are also taken into consideration at this stage. Thus, a reference planning base case is created to represent the "standard" future.

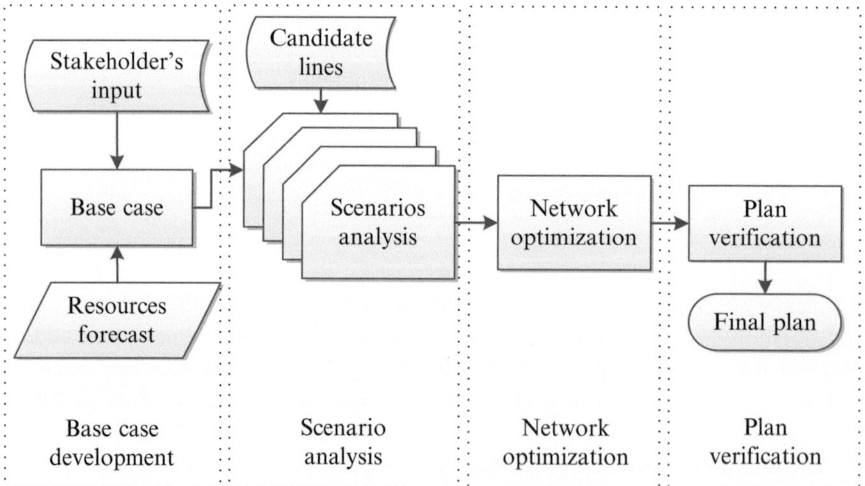

Fig. 6.1 Next-generation transmission expansion planning process

Due to the potential load increase, it is normal to observe some overloads in this reference case and identify where transmission expansion is needed. The base case development is crucial in the TEP process because it serves as a basis for the entire planning framework. After the base case is developed, different scenarios are derived in the next stage based on stakeholders' inputs with the parameters in the base case adjusted to different values. Typical alternative scenarios include combination of variations in loads, energy costs, as well as the generation mix. The candidate lines together with the scenarios serve as inputs to the next stage of network optimization. The network optimization is the core of the entire TEP framework.

Traditionally, due to the lack of efficient algorithms, this step was usually conducted using a trial and error approach. That is, the value of the expansion projects were evaluated by running production cost models with different sets of new lines. However, this approach is by nature a heuristic approach and is only feasible when the candidate line pool is small. As the number of candidate lines increases, the computational burden can easily become intractable. With the development of advanced optimization algorithms, the trend of the next-generation TEP exercise will include production cost analysis based on a mixed-integer programming formulation, which simultaneously optimizes the network expansions and economic dispatch. After the network is optimized for each of the scenarios, transmission lines that appear in most scenarios are viewed as projects with high value and added to the base case.

In the final stage, the adequacy and security (static and dynamic stability) of the expanded system should be evaluated using complete static and dynamic system models with the expectation to identify possible sub-regional reinforcement.

The TEP practice generally consists of two stages: grid optimization and reliability validation. In grid optimization, the TEP model can be viewed as an extension of the optimal power flow (OPF) problem. The TEP and OPF problems share a common basis in the sense that they are both constrained optimization problems. The main difference, however, is that the TEP problem optimizes the network topology based on the economic dispatch. In the grid optimization stage, binary variables are used to determine the status of potential transmission lines and make the TEP formulation "mixed-integer" in nature. In order to obtain an efficient solution, the linearized power flow model and its variations are extensively used in the TEP grid optimization model. For impartial transmission planning organizations such as independent system operators (ISOs), the commonly used objective function is to co-optimize the investment cost and the operating cost over a time horizon.

6.3 Available Software Tools for TEP

Many software tools are available to facilitate analytical studies in power system engineering. In order to perform a systematic planning study, two types of software tools are usually required. The first type of tools is used to perform power system

Table 6.1 Available software tools for power system analysis

Tools name	PF	OPF	Contingency	TS	SS	VS
DSATools [34]	√		√	√	√	√
PowerWorld [37]	√	√[a]	√	√		√
Neplan [38]	√	√	√	√	√	√
PSLF [36]	√		√	√		√
PSS/E [35]	√	√	√	√	√	√
PSAT [39]	√	√	√	√	√	√
MatPower [40]	√	√				

[a]Direct current optimal power flow (DCOPF) and a linearized alternating current optimal power flow (ACOPF)

Table 6.2 Available software tools for planning studies

Tool name	SCED	SCUC	TEP	GEP
GridView	√(DC[a])	√		
PROMOD	√(DC)	√		
UPLAN	√(DC/AC[b])	√		
PSR Net-Plan	√(DC)		√	
PLEXOS	√(DC)	√	√	√

[a]The DC power flow model is used in economic dispatch
[b]The DC or the AC power flow model is used in economic dispatch

studies including power flow, contingency, as well as stability analysis. The commonly used commercial software tools and their functionalities are summarized in Table 6.1. Practitioners should become familiar with one or more of these tools in order to perform the power system studies shown in Table 6.1.

The second type of tools performs planning studies. Their functionality includes security-constrained economic dispatch (SCED), security-constrained unit commitment (SCUC), transmission expansion planning (TEP) as well as generation expansion planning (GEP). Table 6.2 provides a list of planning tools that are commonly used in the industry. GridView [43], PROMOD [44], and UPLAN [45] are used for production cost analysis up to 8,760-h a year. For SCED analysis, UPLAN supports both the direct current (DC) and the alternating current (AC) power flow model. PLEXOS [46] uses mixed-integer based models to perform transmission and generation expansion planning studies. PSR Net-Plan [47] is an integrated computational environment for transmissions analysis and expansion studies. Its module OptNet is specifically designed for TEP analysis. These software packages are widely used in today's power industry in the USA and offer similar functionalities.

Both modeling languages and general high-level languages can be used to formulate the TEP problem. Modeling languages such as AMPL [48], GAMS [49], and AIMMS [50] are tools designed to formulate large and complex optimization models conveniently. Through these languages, users can "describe" the optimization problem by specifying the objective function and constraints; the language will then "translate" the problem into a matrix form and pass it to solvers. However, most of these languages are not free. If budget is a concern, languages such as C/C++, JAVA, Python and MATLAB can also be used to formulate the problem. The advantage of using these languages is that they are easy to access; however, users are responsible for formulating the problem in a matrix form, which could be challenging for large problems.

Depending on the nature of the problem, different solvers can be used. The commonly used commercial solvers include CPLEX [41], Gurobi [42], XPRESS [51] and Knitro (nonlinear solver) [52], free solvers such as CBC [53], SCIP [54], and IPOPT [55] are also available. NEOS server [56] offers a free Internet-based service for solving optimization problems, on which the solvers available represent the state of the art in optimization software.

6.4 TEP Using the Lossless DC Model

The TEP model based on the losses DC power flow model has the following form:

$$\min c_k z_k \tag{6.1}$$

subject to

$$\sum_{k\in\Omega_k^i} P_k + \sum_{g\in\Omega_g^i} \mathrm{PG}_g = \sum_{d\in\Omega_d^i} \mathrm{PD}_d \quad \forall i \in \Omega_b \tag{6.2}$$

$$P_k = -b_k\theta_k \quad \forall k \in \Omega_k \tag{6.3}$$

$$-(1-z_k)\cdot M_k \le (P_k + b_k\theta_k) \le (1-z_k)\cdot M_k \quad \forall k \in \Omega_k^+ \tag{6.4}$$

$$-P_k^{\max} \le P_k \le P_k^{\max} \quad \forall k \in \Omega_k \tag{6.5}$$

$$-z_k P_k^{\max} \le P_k \le z_k P_k^{\max} \quad \forall k \in \Omega_k^+ \tag{6.6}$$

$$0 \le \mathrm{PG}_g \le \mathrm{PG}_g^{\max} \quad \forall g \in \Omega_g. \tag{6.7}$$

where, the objective function (6.1) is to minimize the total investment cost. The nodal balance equation is shown in (6.2), where the net power injection at a bus is equal to the total loads connected to the bus. As shown in (6.3) and (6.4) respectively, the active power flows for existing lines are determined by the product of the line susceptance b_k and the voltage phase angle difference θ_k, while for prospective lines, the big-M method needs to be applied to avoid the presence of nonlinear terms. If a prospective line is selected, i.e., z_k is 1, then (6.4) is forced to be an equality constraint as (6.3), otherwise, z_k is 0, the positive number M_k guarantees that (6.4) is not binding. Constraints (6.5) and (6.6) limit the active power on existing lines and prospective lines respectively. If a prospective line is selected, then (6.6) is the same as (6.5), otherwise, the power flow is forced to be zero. The generator output limit is enforced by (6.7).

The difficulty with the Big-M method is the choice of a proper M. In practice, an arbitrary large M will result in numerical difficulties in the solution by dominating the calculations, however, if M is not large enough, then the true optimal solution will be excluded from the feasible region that causes the branch-and-bound process terminates at only a suboptimal or even with an infeasible solution. As shown in (6.4), in the TEP model, the choice of M_k depends on the parameters of the existing network topology. In order to calculate a proper value of M_k, two situations are discussed: the simple situation is when a candidate line is in an existing transmission corridor. In this case, if the candidate line is not selected, then according to (6.6), $P_k = 0$. As a result, (6.4) can be rewritten as,

$$-M_k \leq b_k\theta_k \leq M_k. \tag{6.8}$$

Considering there are m existing lines in the transmission corridor, the value of M_k can be calculated as,

$$M_k = \min\left(P_{k'}^{\max}/b_{k'}\right)b_k \tag{6.9}$$

where $k' = 1, 2, \ldots, m$, represents all the existing lines in the transmission corridor. When a candidate line creates a new transmission corridor, the problem becomes difficult. According to ref. [4], the shortest path between two terminals of the candidate line needs to be calculated and the computation can be burdensome. In fact, it is not practical to calculate the exact M value for each candidate line that creates a new transmissions corridor, instead, a heuristic upper bound, $2\pi b_k$, is used throughout this chapter.

The above TEP model is tested on Garver's 6-bus system. As shown in Fig. 6.2, the system has six existing lines, five loads, and three generators [3]. Initially, the

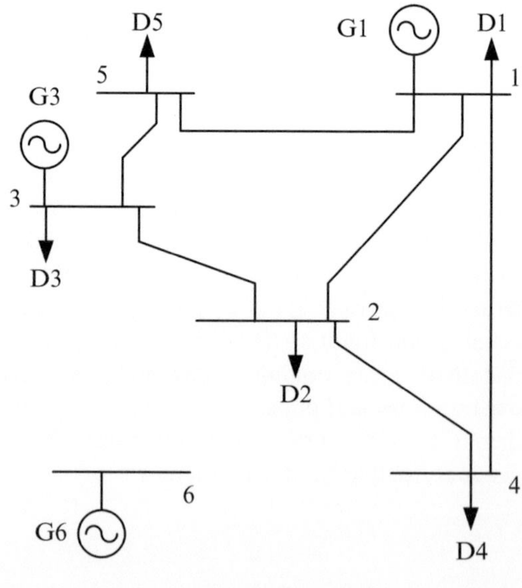

Fig. 6.2 One line diagram of the original Garver's 6-bus system

Table 6.3 Bus data for Garver's 6-bus system

Bus	PG^{min} (MW)	PG^{max} (MW)	Load (MW)
1	0	400	80
2	0	–	240
3	0	400	40
4	0	–	160
5	0	–	240
6	0	600	–

Table 6.4 Branch data for Garver's 6-bus system

Branch	Resistance (p.u.)[a]	Reactance (p.u.)	Cost (10^6 $)	Capacity (MW)
1–2	0.10	0.40	40	100
1–3	0.09	0.38	38	100
1–4	0.15	0.60	60	80
1–5	0.05	0.20	20	100
1–6	0.17	0.68	68	70
2–3	0.05	0.20	20	100
2–4	0.10	0.40	40	100
2–5	0.08	0.31	31	100
2–6	0.08	0.30	30	100
3–4	0.15	0.59	59	82
3–5	0.05	0.20	20	100
3–6	0.12	0.48	48	100
4–5	0.16	0.63	63	75
4–6	0.08	0.30	30	100
5–6	0.15	0.61	61	78

[a]100 MVA base

generator at bus 6 is to be connected to the main system. The data of the system are provided in Tables 6.3 and 6.4. It is assumed that at most three lines are allowed in each transmission corridor. The total number of candidate lines is 39.

In Fig. 6.3, the dashed lines represent new lines to be added. In order to connect bus 6 to the main system and serve the existing loads, four additional lines need to be added. The total investment cost is 110 million dollars (M$).

6.5 A Relaxed ACOPF Model

The approximations made in the traditional DC model significantly simplify the full AC model, but these approximations also degrade the accuracy of the DC model in some cases. In order to improve the model accuracy, the linearized model presented in this section retains a linear representation of reactive power, off-nominal bus voltage magnitudes as well as network losses. The linearization of the line flow

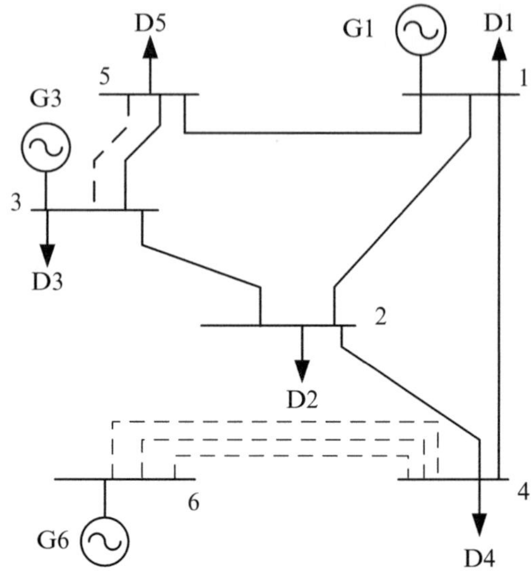

Fig. 6.3 The expanded Garver's 6-bus system

equations is essentially based on a Taylor series and the following assumptions are assumed to be valid:

- The bus voltage magnitudes are always close to 1.0 per unit (p.u.).
- The angle difference across a line is small so that $\sin(\theta_k) \approx \theta_k$ and $\cos(\theta_k) \approx 1$ can be applied. This assumption is valid at the transmission level where the active power flow dominates the apparent power flow in the lines.

6.5.1 *Linearization of the Full AC Model*

If the effects of phase shifters and off-nominal transformer turns ratios are neglected, the AC power flow in branch k between nodes i and j is written as follows,

$$P_k = V_i^2 g_k - V_i V_j (g_k \cos\theta_k + b_k \sin\theta_k) \tag{6.10a}$$

$$Q_k = -V_i^2 (b_k + b_{k0}) + V_i V_j (b_k \cos\theta_k - g_k \sin\theta_k). \tag{6.10b}$$

Based on the first assumption above, the bus voltage magnitude can be written as,

$$V_i = 1 + \Delta V_i \tag{6.11}$$

where $\Delta V^{\min} \le \Delta V_i \le \Delta V^{\max}$ is expected to be small. Substituting (6.11) into (6.10a) and (6.10b) and neglecting higher order terms,

$$P_k \approx (1 + 2\Delta V_i)g_k - \big(1 + \Delta V_i + \Delta V_j\big)(g_k + b_k\theta_k) \tag{6.12a}$$

$$Q_k \approx -(1 + 2\Delta V_i)(b_k + b_{k0}) + \big(1 + \Delta V_i + \Delta V_j\big)(b_k - g_k\theta_k). \tag{6.12b}$$

Notice that (6.12a) and (6.12b) still contain nonlinearities. Since ΔV_i, ΔV_j, and θ_k are expected to be small, the product $\Delta V_i\theta_k$ and $\Delta V_j\theta_k$ can be treated as second order terms and therefore negligible. The linearized power flow equations for line k metered at bus i are obtained as follows,

$$P_k^{ij} = \big(\Delta V_i - \Delta V_j\big)g_k - b_k\theta_k \tag{6.13a}$$

$$Q_k^{ij} = -(1 + 2\Delta V_i)b_{k0} - \big(\Delta V_i - \Delta V_j\big)b_k - g_k\theta_k. \tag{6.13b}$$

The power flow for the same line but metered at bus j is obtained in the same way,

$$P_k^{ji} = -\big(\Delta V_i - \Delta V_j\big)g_k + b_k\theta_k \tag{6.13c}$$

$$Q_k^{ji} = -\big(1 + 2\Delta V_j\big)b_{k0} + \big(\Delta V_i - \Delta V_j\big)b_k + g_k\theta_k. \tag{6.13d}$$

Since P_k and Q_k are linearized, the MVA limit for line k can be written as a second-order cone constraint,

$$P_k^2 + Q_k^2 \leq \big(S_k^{\max}\big)^2. \tag{6.14}$$

Assuming each generator has a quadratic total cost curve,

$$\mathrm{CG}_g = a_g\mathrm{PG}_g^2 + b_g\mathrm{PG}_g + c_g. \tag{6.15}$$

Notice that (6.14) and (6.15) are still convex and can be handled by most commercial linear solvers such as Gurobi.

6.5.2 *Network Losses Modeling*

Unlike the full AC model that inherently captures the network losses, the network losses for the proposed model, however, need to be modeled separately. Using the second order approximation of $\cos\theta_k$ and neglecting high order terms, the network losses can be approximated as,

$$\mathrm{PL}_k \approx g_k\theta_k^2 \tag{6.16a}$$

$$\mathrm{QL}_k \approx -b_k\theta_k^2. \tag{6.16b}$$

Notice that (6.16a) and (6.16b) are non-convex constraints and need to be piecewise linearized. The following MILP formulation is presented to achieve this objective rigorously:

$$\theta_k^2 \approx \sum_{l=1}^{L} k(l)\Delta\theta_k(l) \tag{6.17a}$$

where

$$\theta_k = \theta_k^+ - \theta_k^- \tag{6.17b}$$

$$\sum_{l=1}^{L} \Delta\theta_k(l) = \theta_k^+ + \theta_k^- \tag{6.17c}$$

$$0 \leq \theta_k^+ \leq \delta_k \theta^{\max} \tag{6.17d}$$

$$0 \leq \theta_k^- \leq (1 - \delta_k)\theta^{\max} \tag{6.17e}$$

$$0 \leq \Delta\theta_k(l) \leq \theta^{\max}/L, \quad l = 1, \ldots, L \tag{6.17f}$$

$$\Delta\theta_k(l) \leq \Delta\theta_k(l-1), \quad l = 2, \ldots, L \tag{6.17g}$$

$$\theta^{\max}/L - \Delta\theta_k(l-1) \leq u_k(l-1)\theta^{\max}/L, \quad l = 2, \ldots, L \tag{6.17h}$$

$$\Delta\theta_k(l) \leq [1 - u_k(l-1)]\theta^{\max}/L, \quad l = 2, \ldots, L \tag{6.17i}$$

$$k(l) = (2l - 1)\theta^{\max}/L.$$

In (6.17b), two slack variables $\theta_k{}^+$ and $\theta_k{}^-$ are used to replace θ_k. In (6.17c), the sum of $\theta_k{}^+$ and $\theta_k{}^-$ is used to represent $|\theta_k|$, which is expressed as the summation of a series of linear blocks $\Delta\theta_k(l)$. Constraints (6.17d) and (6.17e) ensure that the right-hand side of (6.17c) equals $|\theta_k|$, while (6.17f)–(6.17i) guarantee that the linear blocks on the left will always be filled up first as illustrated by the shaded area in Fig. 6.4. This MILP formulation eliminates the fictitious losses using binary variables. However, addition of the binary variables tends to complicate the resultant model and makes its efficient solution difficult when the problem scale is large. Alternatively, a relaxed model can be used by excluding (6.17g)–(6.17i) or even (6.17d) and (6.17e) to strike a balance between the computation time and model accuracy.

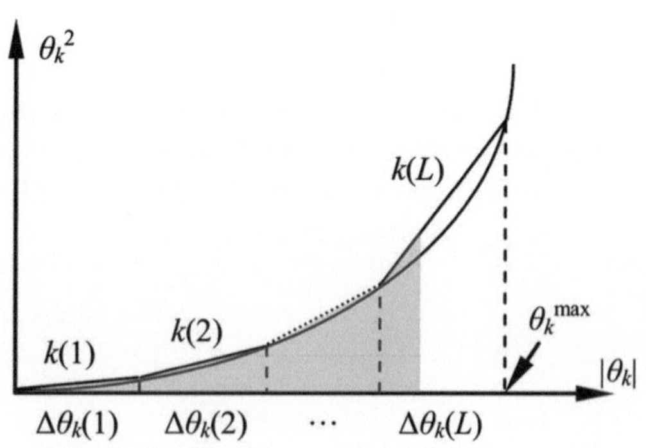

Fig. 6.4 Piecewise linearization of $\theta_k{}^2$

6.6 LACTEP Model

The TEP problem is an extension of the optimal power flow (OPF) problem because it essentially solves a series of OPF problems with different network topologies. In this section, the LACTEP model is developed based on the linearized network model presented in Section 6.5. In this model, it is assumed that the planners have perfect information about the existing network as well as the parameters of the potential lines. Notice that the focus of this chapter is to advance network modeling. Therefore, the planning work is carried out at the peak loading hour for a single future scenario. In real-world applications, however, multiple scenarios can be developed to account for uncertainties and a two-stage stochastic programming planning model can be readily formulated using the LACTEP model proposed in this chapter.

6.6.1 Objective Function

The objective function used in this chapter jointly minimizes the investment cost and the total operating cost,

$$\min C = \sum_{k \in \Omega_k^+} \frac{c_k z_k}{(1+d)^{\mathrm{TP}-1}} + \sum_{t=\mathrm{TP}}^{\mathrm{TP}+\mathrm{TO}} \sum_{g \in \Omega_g} \frac{8,760\,\mathrm{CF}_{gt}\,\mathrm{CG}_{gt}}{10^6(1+d)^{t-1}} \tag{6.18}$$

In (6.18), the first term represents the line investment cost and the second term corresponds to the total operating cost over a time horizon scaled by the generator capacity factor, both in M$ and are discounted to the present value. Notice that the scaled operating cost provides only an estimate of the true operating cost, and can be replaced by a more accurate production cost model if the yearly load profile is available. As implied by the planning timeline in Fig. 6.5, all the selected lines are committed in the targeted planning year, and the operating costs are evaluated over multiple years thereafter. In reality, it is difficult to control the choice of the line to be built in a particular year over the planning horizon. Issues such as project review process, construction and the load forecast accuracy could bring too many uncertainties and make the dynamic planning process intractable. This chapter is based

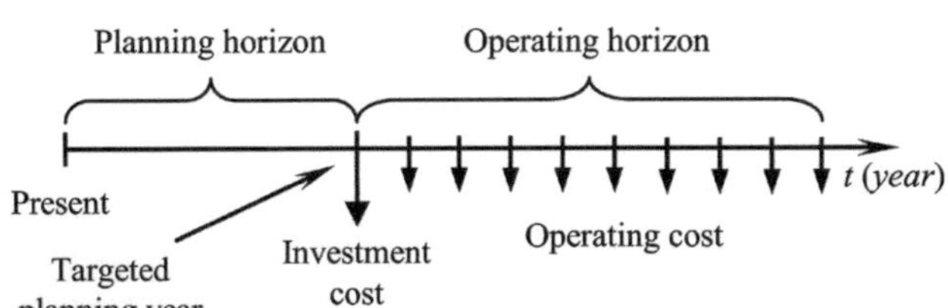

Fig. 6.5 Planning timeline

on a static planning framework and focuses only on the large economic impact of the TEP project. Thus, the incremental economic benefit is lumped into the single targeted planning year.

6.6.2 Power Flow Constraints

In order to build the TEP model, the linearized power flow equations derived in Section 6.5 need to be reformulated. The constraints set related to the power flow equations in the LACTEP model are shown as follows,

$$P_k = p(\Delta V_i, \theta_k) \ \ \forall k \in \Omega_k \tag{6.19a}$$

$$Q_k = q(\Delta V_i, \theta_k) \ \ \forall k \in \Omega_k \tag{6.19b}$$

$$(z_k - 1)M \le P_k - p(\Delta V_i, \theta_k) \le (1 - z_k)M \ \ \forall k \in \Omega_k^+ \tag{6.19c}$$

$$(z_k - 1)M \le Q_k - q(\Delta V_i, \theta_k) \le (1 - z_k)M \ \ \forall k \in \Omega_k^+ \tag{6.19d}$$

$$-z_k S_k \le P_k \le z_k S_k \ \ \forall k \in \Omega_k^+ \tag{6.19e}$$

$$-z_k S_k \le Q_k \le z_k S_k \ \ \forall k \in \Omega_k^+ \tag{6.19f}$$

$$P_k^2 + Q_k^2 \le \left(S_k^{\max}\right)^2 \ \ \forall k \in \Omega_k \cup \Omega_k^+ \tag{6.19g}$$

$$-\theta^{\max} \le \theta_k \le \theta^{\max} \ \ \forall k \in \Omega_k \tag{6.19h}$$

$$(z_k - 1)\pi - \theta^{\max} \le \theta_k \le (1 - z_k)\pi + \theta^{\max} \ \ \forall k \in \Omega_k^+. \tag{6.19i}$$

Constraints (6.19a)–(6.19d) represent the linearized power flow equations for existing lines and prospective lines, where $p(\Delta V_i,\theta_k)$ and $q(\Delta V_i,\theta_k)$ are defined as the right-hand side of (6.13a) and (6.13b) (or (6.13c) and (6.13d)) respectively. For existing lines, the power flow is defined by $p(\Delta V_i,\theta_k)$ and $q(\Delta V_i,\theta_k)$. For prospective lines, the disjunctive constraints (6.19c)–(6.19d) are used to avoid the nonlinearity that would otherwise appear. The power flow on the potential lines is forced to be zero by (6.19e) and (6.19f) if the line is not selected. The line MVA flow is limited by the second-order cone constraint (6.19g). Constraints (6.19h) and (6.19i) put a limit on the phase angle difference across existing lines and prospective lines respectively. If the two buses are directly connected, then θ_k is limited by $\theta^{\max}$ and $-\theta^{\max}$; otherwise, (6.19i) is not binding.

6.6.3 Network Losses

The following constraint set extends the concept of linearized loss modeling to the proposed TEP model,

$$\theta_k = \theta_k^+ - \theta_k^- \quad \forall k \in \Omega_k \cup \Omega_k^+ \tag{6.20a}$$

$$\sum_{l=1}^{L} \Delta\theta_k(l) = \theta_k^+ + \theta_k^- \quad \forall k \in \Omega_k \cup \Omega_k^+ \tag{6.20b}$$

$$0 \le \theta_k^+ \le \delta_k \theta^{\max} \quad \forall k \in \Omega_k \tag{6.20c}$$

$$0 \le \theta_k^- \le (1 - \delta_k)\theta^{\max} \quad \forall k \in \Omega_k \tag{6.20d}$$

$$0 \le \theta_k^+ \le \delta_k \theta^{\max} + (1 - z_k)\pi \quad \forall k \in \Omega_k^+ \tag{6.20e}$$

$$0 \le \theta_k^- \le (1 - \delta_k)\theta^{\max} + (1 - z_k)\pi \quad \forall k \in \Omega_k^+ \tag{6.20f}$$

$$0 \le \Delta\theta_k(l) \le \theta^{\max}/L \quad \forall k \in \Omega_k \tag{6.20g}$$

$$0 \le \Delta\theta_k(l) \le \theta^{\max}/L + (1 - z_k)\pi/L \quad \forall k \in \Omega_k^+ \tag{6.20h}$$

$$\mathrm{PL}_k = g_k \sum_{l=1}^{L} k(l)\Delta\theta_k(l) \quad \forall k \in \Omega_k \tag{6.20i}$$

$$\mathrm{QL}_k = -b_k \sum_{l=1}^{L} k(l)\Delta\theta_k(l) \quad \forall k \in \Omega_k \tag{6.20j}$$

$$0 \le \mathrm{PL}_k \le z_k g_k (\theta^{\max})^2 \quad \forall k \in \Omega_k^+ \tag{6.20k}$$

$$0 \le -\mathrm{PL}_k + g_k \sum_{l=1}^{L} k(l)\Delta\theta_k(l) \le (1 - z_k)M \quad \forall k \in \Omega_k^+ \tag{6.20l}$$

$$0 \le \mathrm{QL}_k \le -z_k b_k (\theta^{\max})^2 \quad \forall k \in \Omega_k^+ \tag{6.20m}$$

$$0 \le -\mathrm{QL}_k - b_k \sum_{l=1}^{L} k(l)\Delta\theta_k(l) \le (1 - z_k)M \quad \forall k \in \Omega_k^+ \tag{6.20n}$$

$$\Delta\theta_k(l) \le \Delta\theta_k(l-1) \quad \forall k \in \Omega_k \cup \Omega_k^+ \tag{6.20o}$$

$$\theta^{\max}/L - \Delta\theta_k(l-1) \le u_k(l-1)\theta^{\max}/L \quad \forall k \in \Omega_k \tag{6.20p}$$

$$z_k\theta^{\max}/L - \Delta\theta_k(l-1) \le u_k(l-1)\theta^{\max}/L \quad \forall k \in \Omega_k^+ \tag{6.20q}$$

$$\Delta\theta_k(l) \le [1 - u_k(l-1)]\theta^{\max}/L \quad \forall k \in \Omega_k \cup \Omega_k^+ \tag{6.20r}$$

$$k(l) = (2l - 1)\theta^{\max}/L \quad \forall k \in \Omega_k \cup \Omega_k^+.$$

Constraints (6.20c)–(6.20f) ensure that the right-hand side of (6.20b) equals $|\theta_k|$ for existing lines and the selected prospective lines respectively. Constraints (6.20g) and (6.20h) determine the upper and lower bound of a linear block $\Delta\theta_k(l)$ for existing lines and prospective lines respectively. For existing lines and the selected prospective lines, $\Delta\theta_k(l)$ is bounded by zero and $\theta^{\max}/L$, otherwise, (6.20h) is not binding. The active and reactive power losses for existing lines are given by (6.20i) and (6.20j) respectively. For prospective lines, the active and reactive power losses are determined by (6.20k)–(6.20l) and (6.20m)–(6.20n) respectively. Constraints (6.20o)–(6.20r) guarantee that the linear blocks on the left will be filled up first.

Constraints (6.20a)–(6.20r) present a full MILP formulation that linearizes the network losses rigorously without generating fictitious losses. Relaxed models can be formed by removing (6.20o)–(6.20r) or even (6.20c)–(6.20f). The linearized line losses are then split in half and attached to the two terminal buses as "virtual demands." The terms corresponding to the network losses are added to the nodal balance equations as follows,

$$\sum_{g \in i} \mathrm{PG}_g + \sum_{k \in i} P_k - \sum_{k \in i} (0.5\mathrm{PL}_k) = \sum_{d \in i} \mathrm{PD}_d \tag{6.20s}$$

$$\sum_{g \in i} \mathrm{QG}_g + \sum_{k \in i} Q_k - \sum_{k \in i} (0.5\mathrm{QL}_k) = \sum_{d \in i} \mathrm{QD}_d. \tag{6.20t}$$

6.6.4 Generator Capacity Limits

In the planning study, all the generators in the system are assumed to be on-line. The generator outputs are limited by their minimum and maximum generating capacities as shown in (6.21a) and (6.21b). Unit commitment is regarded as an operational problem and is therefore not considered in this model. The generator limits are,

$$\mathrm{PG}_g^{\min} \le \mathrm{PG}_g \le \mathrm{PG}_g^{\max} \ \forall g \in \Omega_g \tag{6.21a}$$

$$\mathrm{QG}_g^{\min} \le \mathrm{QG}_g \le \mathrm{QG}_g^{\max} \ \forall g \in \Omega_g \tag{6.21b}$$

The complete LACTEP model is described by (6.18)–(6.21a, b).

6.6.5 $\mathrm{N} - 1$ Modeling

The computational burden is a major concern in MIP problems. Typically, increase the number of binary variables could potentially slow the solution process. Therefore, the candidate line set should be carefully selected and only the applicable transmission corridors should be included. With a large-scale MIP problem, the solver may have trouble finding an initial feasible solution. In this case, providing a feasible starting point will help reduce the overall simulation time.

The $N - 1$ contingency modeling is another major source of the computational burden. In fact, a complete $N - 1$ analysis in the TEP model for a well-designed power system is generally unnecessary because the number of contingencies that will cause serious overloads is generally limited. The $N - 1$ modeling approach used in [11] was to explicitly invoke the set of network constraints for all possible operating conditions and satisfy all the constraints when solving the optimization problem. However, the model presented in this chapter is more complicated. If the approach in [11] were used, the size of the problem could easily become too large to

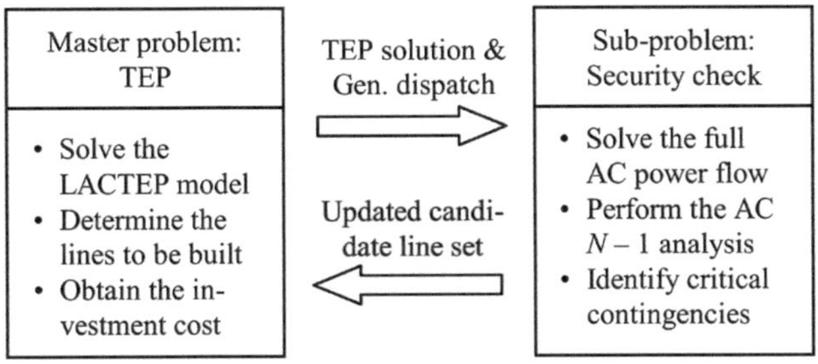

Fig. 6.6 The iterative approach for the $N-1$ contingency modeling

be solvable. Moreover, the TEP problem uses only a relaxed network model, which means that the solution that satisfies the $N-1$ criterion in the TEP model may not represent the actual case in the AC network. In order to make the planned system comply with the $N-1$ criterion without imposing too much computational burden, an iterative approach is proposed in Fig. 6.6.

Using this approach in Fig. 6.6, the original problem is decomposed into a *master problem*, which solves the optimization model and a *sub-problem*, which verifies the network security. The master problem passes the TEP solution and the generator dispatch to the sub-problem, while the sub-problem passes the network violations back to the master problem. The approach solves the two problems iteratively until there is no violation or all the violations identified in the sub-problem are within preset limits.

6.7 Illustrative Examples

In this section, Garver's 6-bus system and the IEEE 118-bus system are studied and the simulation results are demonstrated. The work presented in this chapter is programmed using AMPL. The DC lossless, DC lossy and the LACTEP models are solved by Gurobi. The ACTEP models are solved by Knitro. PowerWorld is used for AC power flow and the $N-1$ contingency analysis. All simulations are done on a Linux workstation with an Intel *i*7-2600, 4-core CPU @ 3.40 GHz with 16 GB of RAM.

6.7.1 Garver's 6-Bus System

Garver's 6-bus system has six existing lines, five loads, and three generators [3]. Initially, the generator connected at bus 6 is isolated from the main system. The system parameters are listed in Tables 6.5 and 6.6. It is assumed that at most three

Table 6.5 Candidate line data for Garver's 6-bus system

Corridor	r_k (p.u.)	x_k (p.u.)	Capacity (MW)	Cost (M$)
1–2	0.04	0.4	100	40
1–3	0.038	0.38	100	38
1–4	0.06	0.6	80	60
1–5	0.02	0.2	100	20
1–6	0.068	0.68	70	68
2–3	0.02	0.2	100	20
2–4	0.04	0.4	100	40
2–5	0.031	0.31	100	31
2–6	0.03	0.3	100	30
3–4	0.059	0.59	82	59
3–5	0.02	0.2	100	20
3–6	0.048	0.48	100	48
4–5	0.063	0.63	75	63
4–6	0.03	0.3	100	30
5–6	0.061	0.61	78	61

Table 6.6 Generator and load data for Garver's 6-bus system

	Load parameters		Generator parameters			
Bus no.	PD (MW)	QD (MVAr)	PG^{min} (MW)	PG^{max} (MW)	QG^{min} (MVAr)	QG^{max} (MVAr)
1	80	16	0	160	−10	65
2	240	48				
3	40	8	0	360	−10	150
4	160	32				
5	240	48				
6			0	610	−10	200

lines are allowed in each transmission corridor. The total number of candidate lines is 39. The objective function is to minimize the line investment cost *only*. The bus voltage magnitude range is 1.00–1.05 p.u. The following two cases are analyzed: *Case 1*: Compare the TEP solutions given by the LACTEP model and other existing models.

Case 2: Network losses sensitivity analysis.

Case 1: In this case, the TEP solution obtained from the LACTEP model is compared with the solutions obtained from other available TEP models. The full MILP approach is used for modeling the network losses. The number of linear blocks is 7. The comparison results are shown in Table 6.7.

The two DC-based TEP models in Table 6.7 seem to be superior in the sense that the investment costs are less. However, the reactive power needed for these two models in the AC network actually exceeds the amount that the three generators can supply. In order to make the AC power flow converge, an additional 189 MVArs and 129 MVArs are needed for the DC lossless and the DC lossy model

Table 6.7 TEP results comparison of Garver's system

TEP model	Expansion plan	Investment cost (M$)	Comments
DC lossless	(3–5), (4–6) × 3	110	Need additional reactive power to make the AC power flow converge. Overloads and undervoltage issues are detected.
DC lossy	(2–6) × 3, (3–5) × 2	130	
LACTEP	(2–3), (2–6) × 2, (3–5) × 2, (4–6) × 3	210	No additional reactive power needed. All indices are within limits.
ACTEP [3]	(2–6) × 3, (2–3) (3–5) × 2, (4–6) × 3, (2–5) × 2	302[a]	

[a]The ACTEP is a non-convex global optimization problem. The result shown in the table is the best solution after 5,000 restarts

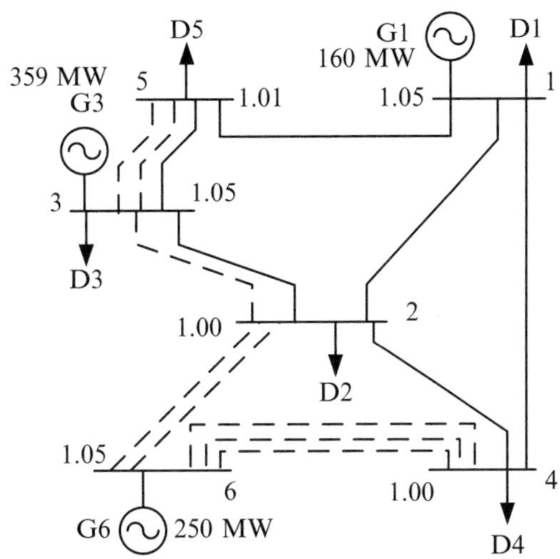

Fig. 6.7 The TEP results of Garver's 6-bus system

respectively. Meanwhile, overloads and under voltage issues are observed in the system, which require additional investment for network reinforcement. The solution obtained from the LACTEP model requires building more lines than the DC-based models do, but needs no additional reactive power and there are no overloads and undervoltage problems in the AC power flow. The expanded Garver's 6-bus system with all indices within the preset limits is plotted in Fig. 6.7.

As a non-convex global optimization problem, multiple starting points are tried to obtain a good solution for the ACTEP model. As shown in Table 6.7, the best objective value for the ACTEP model after 5,000 restarts is still much higher than the objective function given by the LACTEP model. It will also be computationally too expensive to apply the ACTEP model to larger power system planning problems. This comparison reveals that the solutions given by the DC-based TEP models may not represent the actual case in the AC network and additional network reinforcement is likely to be needed. The LACTEP model better approximates the AC network and therefore provides a more realistic TEP solution.

For small systems such as the 6-bus example, reactive power can be a critical issue to make the AC power flow converge. As indicated by Table 6.7, the LACTEP model chooses to build more lines to provide reactive power support. In reality, increasing generator reactive power capacity and installing VAr support devices can certainly be considered as alternative solutions if a DC-based TEP solution is adopted, but one should be aware that it may not be easy to increase reactive power capacity of existing generators, and can be costly to install VAr support devices at high voltage buses, too. For real-world applications, different solution options can be compared to find the most cost-effective TEP plan. For large systems with meshed topology, using LACTEP model is more appropriate because it dispatches generators more accurately, gives a better estimation of line flows, and therefore provides a realistic TEP solution, which DC-based models usually fail to do.

Case 2: As discussed in Section 6.5.2, the linearized network losses can be rigorously modeled using the MILP formulation. However, addition of the binary variables also increases the complexity of the TEP model. The number of linear blocks can significantly affect the solution time as well as the model accuracy. Table 6.8 shows how the number of linear blocks changes the size of the problem and the TEP solution. The full MILP formulation is used for the results shown in Table 6.8.

The variable types in Table 6.8 show that the size of the problem increases as the number of linear blocks increases. This behavior coincides with the intuition that more variables are needed to model the additional linear blocks. It should be noted that the linearization intrinsically overestimates the losses in the system. If too few linear blocks are used, then the overestimation can be significant and the problem will be infeasible with the given set of candidate line set. This is reflected from both

Table 6.8 The effects of number of linear blocks

	Variable types				
Linear blocks	Continuous	Binary	Objective (M$)	Total P losses (MW)	Time (s)
1	281	84	Infeasible	–	–
2	323	126	378	16.3	>413
3	407	171	259	11.8	69
4	449	213	230	8.8	97
5	489	253	230	8.7	33
6	579	298	230	8.2	89
7	621	340	210	8.2	34
8	666	385	210	8.2	43
9	708	427	210	8.2	116
10	748	467	210	8.2	97

Table 6.9 Comparison of different network losses models

Losses modeling approach[a]	Total *P* losses (MW)	Objective (M$)	Time (s)/Δ (%)
Full MILP	8.2	210	34/(0 %)
Relaxation 1 (R1)	10.6	210	20/(−41 %)
Relaxation 2 (R2)	8.4	210	22/(−35 %)
Relaxation 3 (R3)	10.7	210	21/(−38 %)
Do not model losses[b]	0	150	3/(−91 %)

[a]Full MILP: Use (6.20a)–(6.20r) to model the linearized network losses
Relaxation 1: Remove (6.20o)–(6.20r)
Relaxation 2: Remove (6.20c)–(6.20f)
Relaxation 3: Remove (6.20c)–(6.20f) and (6.20o)–(6.20r)
[b]Losses are not modeled, but r_k, Q, and V are retained

the trends of losses and the objective values listed in Table 6.8. It is worth noticing that due to the mixed-integer nature of the problem, the change in solution time does not follow a linear pattern. When too few linear blocks are used, the TEP results may contain unnecessary lines due to the significant overestimation of the network losses. It may also take the solver a long time to branch out an initial feasible solution. On the other hand, too many linear blocks will impose unnecessary computational burden and slow the solution time. The key idea of the study is to find the number of linear blocks that gives the best balance between the model accuracy and the solution time. In this case, 7 is an appropriate number.

The results contained in Table 6.9 compare the accuracy of the relaxed losses models and the solution time. The number of linear blocks used for this study is 7.

Among all the loss modeling approaches listed in Table 6.9, the full MILP formulation is the most accurate and serves as a basis of the study. The R1 approach relaxes the constraints for prioritizing the lower linear blocks. This approach reduces the solution time by approximately 41 %, but the drawback is that it creates 2.4 MW fictitious active power losses. The R2 approach relaxes the constraints for modeling the absolute value. It reduces the solution time by approximately 35 %, and creates only 0.2 MW fictitious losses. The R3 approach relaxes both the constraints that were relaxed in R1 and R2. It reduces the solution time by approximately 38 %, but creates 2.5 MW fictitious losses. Additionally, if losses are ignored, the solution time will be significantly reduced by 91 %, but the TEP solution no longer satisfies preset the voltage requirement. Except for the no loss case, the TEP solutions remain the same for all other loss modeling approaches. One explanation is that the impact of fictitious losses is not significant enough to change the TEP results in this case. The study results show that the R2 approach is considered as the best trade-off between model accuracy and solution time.

6.7.2 *IEEE 118-Bus System*

The IEEE 118-bus system [57] is used to demonstrate the potential of applying the proposed LACTEP model to large power systems. The system has 186 existing

Table 6.10 Zonal data of the IEEE 118-bus system

	Bus	Branch	Generation (MW)	Load (MW)
Zone 1	42	62	2,280	1,865
Zone 2	48	81	4,160	3,125
Zone 3	28	43	2,544	1,271
Total	118	186	8,884	6,261

Table 6.11 TEP planning criterion for the IEEE 118-bus system

	Normal ($N-0$)	Contingency ($N-1$)
Voltage (p.u.)	$0.96 \leq V \leq 1.06$	$0.92 \leq V \leq 1.06$
Power flow	$P_k^2 + Q_k^2 \leq (S_k^{\max})^2$	$P_k^2 + Q_k^2 \leq (1.1 S_k^{\max})^2$

branches, 54 generators, and 91 loads. The line ratings are reduced to create congestions. The zonal data is listed in Table 6.10. The load assumed is the peak loading level. The discount rate is assumed to be 10 %, and the number of linear blocks used for loss modeling is 10. The planning horizon is 10 years. The objective function jointly minimizes the line investment cost and the scaled 10-year total operating cost. The average capacity factors published in [58] are used in this chapter. The capital costs of transmission lines are assumed to be proportional to the length of the lines. Due to the absence of real data, all prospective lines are assumed to share the same corridor and have the same parameters as the existing lines. The planning criteria are given in Table 6.11. The detailed planning procedure is described in the following steps.

Step 1: Run a regular AC power flow on the system to be planned, and identify the lines that are overloaded or heavily loaded. These lines will form the initial candidate line set.

Step 2: Use the candidate line set and run the LACTEP model. Obtain the TEP solution and update the system.

Step 3: Rerun a regular AC power flow on the expanded system and identify any overloaded lines/transformers. Notice that it is still possible to observe some violations in this step because the network model used in the TEP problem is essentially a relaxation of the AC network model. If this happens, one should slightly reduce the line ratings used in the TEP problem and redo Step 2 to Step 3. If no violation is identified in this step, then proceed to Step 4.

Step 4: Perform a complete $N-1$ analysis on the expanded system. Identify the worst contingency and take the line out of service. Form a new candidate line set and return to Step 2. Do this iteratively until all violations are within the preset threshold (as specified in Table 6.11). It is assumed that the generator dispatch do not change during this process.

The flowchart of the iterative approach is plotted in Fig. 6.8.

Table 6.12 shows the 15 initial candidate lines and their cost data. The candidate lines for the $N-1$ contingency analysis are not included in the table.

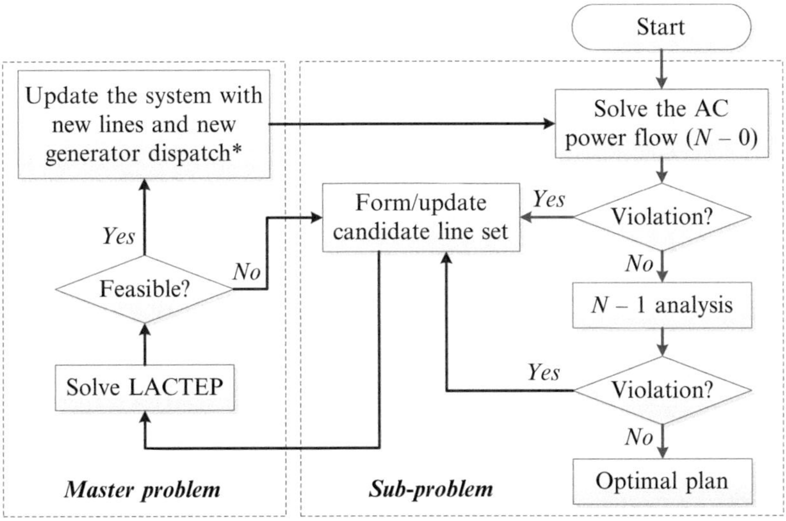

*The updated generator dispatch is only calculated for $N-0$. For $N-1$ analysis, it is assumed that the generator dispatch is fixed.

Fig. 6.8 Flowchart of the iterative approach for considering $N-1$ contingency

Table 6.12 Initial candidate lines for the IEEE 118-bus system

No.	Lines	Cost (M$)	No.	Lines	Cost (M$)
1	(3–5)	16.2	9	(38–37)	6.8
2	(5–6)	9.7	10	(69–67)	15.2
3	(8–9)	5.5	11	(77–78)	2.8
4	(8–5)	6.0	12	(80–99)	30.9
5	(9–10)	5.8	13	(82–83)	6.6
6	(17–113)	5.4	14	(94–100)	10.4
7	(23–32)	17.3	15	(99–100)	14.6
8	(26–30)	15.5			

Table 6.13 The TEP results for $N-0$

Lines to be built	(3–5), (8–9), (9–10), (26–30)
Investment cost (M$)	43
Total operating cost (M$)	1,567.4 (10-year)
Solution time (s)	4

The cost of building a transmission line can be roughly estimated by its length, cost per mile and the cost multipliers [59]. Assuming all lines are 230 kV double circuit lines, then the capital cost of a transmission line is calculated as,

$$C_{\text{line}} = 1.5\beta(\text{Line length}) \tag{6.22}$$

where 1.5 is cost per mile of 230 kV double circuit lines and β is the transmission length cost multipliers. For lines longer than 10 miles, 3–10 miles and shorter than 3 miles, the β values are 1.0, 1.2 and 1.5 respectively. Notice that (6.22) only gives a

Table 6.14 The iterative planning process for $N-1$

Iterations	Contingency line	Violation type	Lines added
1	(77–78)	Line overloading	(77–78) circuit 2
2	(80–99)		(80–99) circuit 2
3	(25–27)		(23–32)
4	(38–65)		(30–38)
5	(1–3)		(1–3) circuit 2
6	(86–87)		(86–87) circuit 2
7	(64–65)		(64–65) circuit 2
8	(60–61)		(60–61) circuit 2
9	(15–17)		(15–17) circuit 2
10	(12–117)	Loss of loads	(12–117) circuit 2
11	(110–117)		(110–117) circuit 2

rough estimate of the line capital cost, more factors need be included in order to obtain a better estimate. The TEP results are demonstrated in Tables 6.13 and 6.14 for $N-0$ and the $N-1$ contingency case respectively.

It is observed from Table 6.13 that four lines need to be added in order to relieve the overloads in the original system with all lines in service ($N-0$). The investment cost is 43 M$, and the estimated 10-year total operating cost is 1,567.4 M$, which is approximately 156.7 M$ per year. The original system is then expanded using the TEP solution in Table 6.13 and solved using the AC power flow with all indices within the limits. Therefore, with the four lines being added, the system is $N-0$ secure. Meanwhile, it is worth mentioning that the TEP solution given by the DC lossless model requires building no line for this case. However, significant overloads and undervoltage issues are observed in the AC power flow.

In order for the system to comply with the $N-1$ criterion, the planning process needs to proceed to Step 4. In this case, only line (do not include transformers) contingencies are considered. During the contingency, the monitored violations monitored are overloads, loss of loads as well as undervoltages. The iterative planning process is elaborated in Table 6.14.

In Table 6.14, the second column lists the lines that are manually outaged in each iteration. The contingencies in the table are ranked in the order of the severity of overload caused in the system. The line that causes severe overloads and results in a large number of associated overloaded lines will be addressed first. The third column shows the type of the violations and the last column provides the solution to mitigate the potential overloads or loss of loads. After 11 iterations, all indices are within the limits set in Table 6.11 for the contingency case. The system complies with the $N-1$ contingency criterion. Mathematically, this iterative approach does not guarantee an optimal solution, but in terms of the computational burden, this approach attains the same goal more efficiently.

6.8 Summary

This chapter discusses the TEP problem in modern power systems. First, the state of the art of the TEP research is reviewed, and the TEP process and the available software tools are presented. The mathematical models of the TEP problem are then discussed, and a new approach to linearize the full AC network model is proposed. The proposed LACTEP model retains a linear representation of reactive power, off-nominal bus voltage magnitudes and network losses. A MILP formulation for network losses modeling is developed to eliminate fictitious losses. An iterative approach is also presented to incorporate the $N-1$ contingency criterion in TEP problems.

Acknowledgement The author would like to thank the Western Electricity Coordinating Council (WECC) for providing the funds for this research. The author would also like to thank Dr. H. D. Mittelmann for providing the ORION computing platform.

References

1. Western Electricity Coordinating Council (WECC) WECC 10-year regional transmission plan summary. Sept 2011. [Online]. http://www.wecc.biz/library/StudyReport/Documents/Plan_Summary.pdf
2. Latorre G, Cruz RD, Areiza JM, Villegas A (2003) Classification of publications and models on transmission expansion planning. IEEE Trans Power Syst 18(2):938–946
3. Garver LL (1970) Transmission network estimation using linear programming. IEEE Trans Power Apparatus Syst PAS-89:1688–1697
4. Bahiense L, Oliveira GC, Pereira M, Granville S (2001) A mixed integer disjunctive model for transmission network expansion. IEEE Trans Power Syst 16(3):560–565
5. Alguacil N, Motto AL, Conejo AJ (2003) Transmission expansion planning: a mixed-integer LP approach. IEEE Trans Power Syst 18(3):1070–1077
6. de la Torre S, Conejo AJ, Contreras J (2008) Transmission expansion planning in electricity markets. IEEE Trans Power Syst 23(1):238–248
7. Garces LP, Conejo AJ, Garcia-Bertrand R, Romero R (2009) A bilevel approach to transmission expansion planning within a market environment. IEEE Trans Power Syst 24(3):1513–1522
8. Khodaei A, Shahidehpour M, Kamalinia S (2010) Transmission switching in expansion planning. IEEE Trans Power Syst 25(3):1722–1733
9. Seifu A, Salon S, List G (1989) Optimization of transmission line planning including security constraints. IEEE Trans Power Syst 4:1507–1513
10. NERC System performance under normal conditions, NERC Standard TPL-001–0.1, Oct 2008
11. Zhang H, Vittal V, Heydt GT, Quintero J (2012) A mixed-integer linear programming approach for multi-stage security-constrained transmission expansion planning. IEEE Trans Power Syst 27(2):1125–1133
12. O'Neill RP, Krall EA, Hedman KW, Oren SS (2012) A model and approach for optimal power systems planning and investment. Math Program 140(2):239–266
13. Hedman KW, O'Neill RP, Fisher EB, Oren SS (2009) Optimal transmission switching with contingency analysis. IEEE Trans Power Syst 24(3):1577–1586

14. Rider MJ, Garcia AV, Romero R (2007) Power system transmission network expansion planning using AC model. IET Gener Transm Distrib 1(5):731–742
15. Zhang H, Heydt GT, Vittal V, Mittelmann HD (2012) Transmission expansion planning using an AC model: formulations and possible relaxations. IEEE PES General Meeting, July 2012
16. Otiveira GC, Costa APC, Binato S (1995) Large scale transmission network planning using optimization and heuristic techniques. IEEE Trans Power Syst 10(4):1828–1834
17. da Silva EL, Ortiz JMA, de Oliveira GC, Binato S (2001) Transmission network expansion planning under a tabu search approach. IEEE Trans Power Syst 16(1):62–68
18. da Silva EL, Gil HA, Areiza JM (2000) Transmission network expansion planning under an improved genetic algorithm. IEEE Trans Power Syst 15(3):1168–1175
19. Binato S, de Oliveira GC, de Araújo JL (2001) A greedy randomized adaptive search procedure for transmission expansion planning. IEEE Trans Power Syst 16(2):247–253
20. Romero R, Gallego RA, Monticelli A (1996) Transmission system expansion planning by simulated annealing. IEEE Trans Power Syst 11(1):364–369
21. Leeprechanon N, Limsakul P, Pothiya S (2010) Optimal transmission expansion planning using ant colony optimization. J Sust Energy Environ 1:71–76
22. Roh JH, Shahidehpour M, Wu L (2009) Market-based generation and transmission planning with uncertainties. IEEE Trans Power Syst 24(3):1587–1598
23. Maghouli P, Hosseini SH, Buygi MO, Shahidehpour M (2011) A scenario-based multi-objective model for multi-stage transmission expansion planning. IEEE Trans Power Syst 26(1):470–478
24. Gorenstin BG, Campodonico NM, Costa JP, Pereira MVF (1993) Power system expansion planning under uncertainty. IEEE Trans Power Syst 8(1):129–136
25. da Silva AML, Ribeiro SMP, Arienti VL, Allan RN, Filho MBDC (1990) Probabilistic load flow techniques applied to power system expansion planning. IEEE Trans Power Syst 5(4):1047–1053
26. Yu H, Chung CY, Wong KP, Zhang JH (2009) A chance constrained transmission network expansion planning method with consideration of load and wind farm uncertainties. IEEE Trans Power Syst 24(3):1568–1576
27. Li W, Choudhury P (2007) Probabilistic transmission planning". IEEE Power Energy Mag 5(5):46–53
28. Choi J, Tran T, El-Keib AA, Thomas R, Oh H, Billinton R (2005) A method for transmission system expansion planning considering probabilistic reliability criteria. IEEE Trans Power Syst 20(3):1606–1615
29. Sánchez-Martín P, Ramos A, Alonso JF (2005) Probabilistic midterm transmission planning in a liberalized market. IEEE Trans Power Syst 20(4):2135–2142
30. Escobar AH, Gallego RA, Romero R (2004) Multistage and coordinated planning of the expansion of transmission systems. IEEE Trans Power Syst 19(2):735–744
31. Braga ASD, Saraiva JT (2005) A multiyear dynamic approach for transmission expansion planning and long-term marginal costs computation. IEEE Trans Power Syst 20(3):1631–1639
32. Xie M, Zhong J, Wu FF (2007) Multiyear transmission expansion planning using ordinal optimization. IEEE Trans Power Syst 22(4):1420–1428
33. Sepasian MS, Seifi H, Foroud AA, Hatami AR (2009) A multiyear security constrained hybrid generation-transmission expansion planning algorithm including fuel supply costs. IEEE Trans Power Syst 24(3):1609–1618
34. PowerTech (2012) Dynamic security assessment software (DSATools) manual, ver. 12
35. Siemens Power Technologies International (PTI) (2013) PSS®E application program interface, ver. 33.4
36. GE Concorda (2012) PSLF user's manual, ver 18
37. PowerWorld Simulator 16 [Online]. http://www.powerworld.com/
38. BCP Switzerland NEPLAN [Online]. Available:http://www.neplan.ch/html/e/e_home.htm
39. Milano F (2005) An open source power system analysis toolbox. IEEE Trans Power Syst 20 (3):1199–1206

40. Zimmerman RD, Murillo-Sánchez CE, Thomas RJ (2011) MATPOWER: steady-state operations, planning and analysis tools for power systems research and education. IEEE Trans Power Syst 26(1):12–19
41. CPLEX Optimizer [Online]. http://www-01.ibm.com/software/commerce/optimization/cplex-optimizer/
42. GUROBI optimization [Online]. http://www.gurobi.com/
43. ABB. GridView—an analytic tool for market simulation & asset performance evaluations [Online]. http://www.abb.com
44. Ventyx PROMOD IV [Online]. http://www.ventyx.com/analytics/promod.asp
45. UPLAN Network Power Model [Online]. http://www.energyonline.com/products/uplane.aspx
46. PLEXOS [Online]. http://energyexemplar.com/products/plexosdesktopedition
47. PSR NetPlan [Online]. http://www.psr-inc.com.br/portal/psr/
48. A Modeling Language for Mathematical Programming (AMPL) [Online]. http://www.ampl.com/
49. General Algebraic Modeling System (GAMS) [Online]. http://www.gams.com/
50. Advanced Interactive Multidimensional Modeling System (AIMMS) [Online]. http://www.aimms.com
51. FICO® Xpress Optimization Suite [Online]. http://www.fico.com/
52. KNITRO [Online]. http://www.ziena.com/knitro.htm
53. Coin-or branch and cut (Cbc) [Online]. https://projects.coin-or.org/Cbc
54. Solving Constraint Integer Programs [Online]. http://scip.zib.de/
55. Interior Point OPTimizer (IPOPT) [Online]. https://projects.coin-or.org/Ipopt
56. Network-Enabled Optimization System (NEOS) Server [Online]. http://www.neos-server.org/neos/
57. Electrical and Computer Engineering Department, Illinois Institute of Technology (IIT). IEEE 118-bus system data [Online]. http://motor.ece.iit.edu/Data/Gastranssmion_118_14test.xls
58. U.S. Energy Information Administration. Electric power annual 2009, April 2011 [Online]. http://www.eia.gov/electricity/annual
59. Tim Mason, Trevor Curry, Dan Wilson (2012) Capital costs for transmission and substations: recommendations for WECC transmission expansion planning, Oct 2012 [Online]. https://www.wecc.biz/committees/BOD/TEPPC/12110102/Lists/Minutes/1/BV_WECC_TransCostReport_Final.pdf

Chapter 7
Evolution of Smart Distribution Systems

Anil Pahwa

7.1 Introduction

Power distribution systems are at the lowest end of the power grid and thus are nearest to the customers. It is estimated that capital invested in power distribution systems worldwide is 40 % of the total investment in power systems. Of the remaining 60 %, generation accounts for 40 % and transmission accounts for 20 %. Customers experience direct impact of events occurring in distribution systems because they are directly connected to them. According to some reports, 80 % of the interruptions experienced by customers are due to outages in distribution systems [1] and on an average a failure of a segment on a feeder will interrupt service to about half of the customers it serves. Although power distribution systems are a large part of power systems and have a direct impact on the customers, integration of automation into their operation and control has lagged considerably behind those of generation and transmission systems. Progress on power distribution system automation has been relatively slow due to the large investments needed to automate these systems, with an extremely large number of components. Now with infusion of Smart Grid [2–4] technology, new challenges and opportunities are emerging. Smart Grid initiatives and funding by the federal government to utilities for implementing Smart Grid technologies has accelerated activities related to distribution automation and smart metering. Similarly, the number of customers installing rooftop solar generation or owning plug-in hybrid or electric vehicles is increasing gradually. High penetration of such devices creates new dynamics for which the current equipment in distribution system is inadequate. Rapid fluctuations of power output from distributed renewable resources cause voltage control problems [5], which requires new approaches to operate distribution

A. Pahwa (✉)
Electrical and Computer Engineering, Kansas State University, Manhattan, KS, USA
e-mail: pahwa@ksu.edu

E. Kyriakides et al. (eds.), *Electric Power Engineering Research and Education*,
Power Electronics and Power Systems, DOI 10.1007/978-3-319-17190-6_7

systems. Similarly, new technologies are needed to permit operation of a distribution system as a microgrid with high penetration of renewable resources. Such advances will be of extreme value to maintain availability of power supply to customers upon loss of power from the grid and under natural disasters, such as hurricanes or earthquakes, and terrorist acts [6].

Although distribution systems are a significant part of the power systems, very little real-time information is available to operators from the system at this level. Most often the only real-time measurement available for distribution systems is from the feeder gateway at the substation. As a result most of the operation and planning of distribution systems has relied on heuristics and archived information. Due to lack of automation, most of the distribution systems operate in non-optimum mode and have difficulties in recovering from abnormal events. Attempts to automate electricity distribution to improve system operation have been ongoing since the introduction of the concept of Distribution Automation (DA) in the 1970s. Advances in computer and communication technology have made distributed automation possible. Automation allows utilities to implement flexible control, which would result in enhanced efficiency, reliability, and quality of electric service. Flexible control also results in more effective utilization and life-extension of the existing distribution system infrastructure. Several utilities have run pilot projects and some have implemented automation based on their needs. However, there are no cases where we find comprehensive automation of distribution systems. In parallel with distribution automation, significant activity has taken place in the Automated Metering Infrastructure (AMI), which deals mainly with the placement of smart meters in homes to measure and monitor electricity, gas, and water consumption. Information from AMI systems has also been used by utilities for outage management.

Now, with additional progress in technology, the current level of automation is not sufficient. Until now, the major focus of Smart Grid has been on advanced metering but now the utilities are focusing on distribution automation. Distribution systems of the future will have homes with smart meters to monitor energy consumption, on-site grid-connected solar or wind generation, battery storage, and plug-in vehicles. The feeders will have advanced power electronic switching devices to control the system, sensors at strategic locations to measure flow of real and reactive power, voltage, and current. Similarly, the substation will have power electronic controls, measurements, and protection to operate the system more efficiently and reliably. The system will have a seamless communication layer from the utility's control room to customers and it will be integrated with advanced cyber systems to enable its operation. Substantially more real-time information will be available to facilitate their operation and control [4].

In this chapter the evolution of distribution systems from manual to smart automated systems is examined. Various contemporary issues that are relevant for the modern distribution systems are also discussed.

7.2 Important Issues for Distribution Systems

7.2.1 *Reliability*

Distribution systems typically have operating voltages lower than 35 kV and feeder lengths range from 1 to 10 miles with some feeders longer than that in rural systems. A large part of the distribution systems in the USA is overhead with radial configuration for economic and technical reasons. Underground feeders cost five to ten times more than the overhead feeders. On the technical side, protection of a radial system is simpler. This practice has been followed for over a century and still continues to be the prominent mode. Only distribution systems in downtowns or business districts in cities are mostly underground with network configuration. The main motivation for this practice is aesthetics and to provide higher reliability of power supply to businesses to prevent slowdown of economic activity. Now, many cities and towns are requiring underground systems in residential areas for esthetics. This does not imply that the main feeders are underground, only the last part of the system closest to the customers becomes underground.

Since a large part of life in the modern society depends on electricity, any interruption, momentary or sustained, results in great inconvenience and financial loss. The utility companies strive to maintain a certain level of reliability in the generation, transmission, and distribution part of the power system. The reliability of a distribution system correlates directly with its ability to deliver power to the customers without failure. Although historically utilities have maintained a very high level of reliability, pressure on them to continue to maintain this has gradually increased over the past several years since some state utility commissions are imposing or proposing penalties on utilities for not providing certain expected levels of reliability. The situation is further compounded by the fact that customers of the digital age are expecting a higher level of reliability and the utilities are operating under a tighter budget. Thus, distribution system reliability is becoming a very significant part of the utility business [7]. With automation, utilities are able to detect outages sooner as well as restore the system sooner, thus improving reliability.

7.2.2 *Power Quality*

Another issue important to utilities and customers is power quality. Over the years due to increased penetration of computers, other electronics and power electronics devices, distribution systems have experienced a high level of harmonics. Although most residential customers do not experience any direct problems with harmonics, industrial customers may experience some problems due to higher level of harmonics. Moreover, harmonics cause higher losses in the system and thus utilities are always seeking solutions to mitigate the level of harmonics in the system. In addition, voltage sag and surge are important for operation of computers and computerized equipment.

Extended sag will result in shut down of computers and excessive surge can cause damage to the equipment. Since it is impossible to eliminate voltage sags and surges completely, computer equipment manufacturers usually follow the CBEMA (Computer and Business Equipment Manufacturer's Association) or ITIC (Information Technology Industry Council) recommended curves for manufacturing their equipment [8]. All the equipment should operate normally within this envelope for voltage deviation and power continuity. Real-time monitoring allows utilities to take corrective actions to mitigate power quality concerns.

7.2.3 Efficiency

In addition to reliability and quality, utilities are also concerned about efficiency of distribution systems. Efficiency can be defined both in terms of manpower utilization and in terms of technical efficiency. Technical efficiency refers to reduction in losses by proper design and operation of the system. For example, proper selection of conductors for the feeders reduces system losses. Implementing system reconfiguration and Volt/VAr control using automated equipment can reduce losses significantly. Further, automation allows for more efficient utilization of manpower.

7.2.4 Demand Response

In conjunction with Smart Grid activities, utilities are also contemplating time differentiated pricing for electricity. Currently, in most situations, residential customers pay a fixed rate for electricity at all times. In the new paradigm, the rate will be higher during the peak hours in the peak season. For example, for many utilities this might mean higher rates from early afternoon to early evening in 3 or 4 months of summer. To compensate for increase in these hours, the rate will be dropped for other hours. Other options are Critical Peal Pricing (CPP) and Real-Time Pricing (RTP) [9]. CPP entails very high rates (up to 10 times the normal rate) during the peak hours on 10–15 days in the year. As the name implies, RTP means real-time hourly price for electricity for 10–15 critical days of the year. In both cases, the customers are informed a day-ahead of these rates to allow customers to react to them. The idea behind these rates is to solicit demand response from customers to reduce system load on critical days. Reduction in load allows utilities to save by not generating electricity by using the expensive peaking units or by not purchasing expensive electricity in the spot market. Customers, on the other hand, save on their yearly electricity bill. Therefore, it is a win–win situation for both customers and utilities. Time-of-use (TOU) rates and direct control of appliance has been implemented in the past, particularly in the early days of demand-side management in the 1980s, with limited success. Now, with newer technologies and with innovative rates, such as CPP and RTP, more impact on demand response is expected.

7.2.5 Distributed Generation

Distribution systems typically do not have generation. Due to economy of scale historically generation has been at the bulk level. However, over the past 15 years some types of generation have become viable at the distribution level. These generators connect to the distribution system at the primary voltage level. These generators create significant protection and operation problems in the system by altering the direction of power flows in systems that are primarily designed for radial operation. Integration of these resources has been a significant challenge. IEEE has been working to develop standards to overcome these challenges [10]. Now, with increased emphasis on renewable energy, individual customers are getting interested in owning a rooftop solar generation or a small wind mill. These resources have to be connected to the utility system at the secondary voltage level. As penetration of such customer-owned generation increases, new and unknown challenges are emerging. Advanced automation of the system is needed to handle such situations [11].

7.2.6 Integrated System Planning

In a simple sense distribution system planning means being prepared for the future to meet the electricity needs of the customers. Since the loads change continuously due to change in population, change in technology, and change in habits of people, planning for electricity needs becomes a very complex and capital-intensive process. It requires consideration of several objectives simultaneously while meeting the technical constraints. The overall goal is to minimize the total cost while ensuring that the system has adequate capacity to supply the load in the future with adequate reliability and acceptable voltage quality. Good plans should be able to address the short-term needs of 1–2 years as well as long-term needs ranging from 5 to 10 years.

The first stage in planning is to forecast the load in the future. In addition to knowing the extent of load growth, it is important to know where the load will be growing. After the areas of load growth and future load levels are known, the next step is the determination of location and capacity of substations. Details include number and size of transformers. The planning results may include upgrading the existing substations as well as building new substations. The final step is feeder design to deliver power from substations to the customers. Feeder design includes both the primary and the secondary systems. The decisions include number of primary feeders and size of conductors, and their routing.

At the secondary level, the decisions include location and size of distribution transformers. Most of the secondary constructions in the USA have been overhead so far. However, now there is a trend to make all new secondary construction underground. Although underground secondary system can be five to ten times more expensive than overhead secondary, underground secondary is preferred by many localities due to aesthetics and for greater reliability.

Legacy distribution systems have very little metering information available from the system. Usually, loads on the substation transformers and the feeders are measured. Thus, utilities have very little information beyond the feeder. Utilities typically rely on customer billing data and the total installed distribution transformer capacity to get an estimate of the existing load. As part of the load research activity, utilities also installed recording devices at selected customer locations to record loads at a predetermined interval (5-min, 15-min, 30-min, or 1-h) to get an idea of daily load profiles for different classes of customers. These load profiles are also useful for planning purposes. However, with limited information available from the system, distribution planning was ad hoc, which would often result in over designed systems.

Modern distribution systems have significantly more metering capabilities, both at the system level due to distribution automation and at the customer-level due to AMI. Although many advances in automation and metering have been made over the years, by and large the existing distribution systems around the world do not have much metering capabilities. As more systems become automated, more accurate data from different parts of the system becomes available to the distribution planners. Such availability of data removes the need to make assumptions about various factors about which no data of information were available.

Traditionally, distribution systems are designed to meet peak demand. The modern approach based on risk analysis requires number of hours spent at different load levels. These data allow the planners to determine the risk of insufficient capacity to meet the demand over a period of time. Specifically, number of hours under peak conditions is very important for risk assessment. Modern distribution planning must be integrated with distribution automation and it must consider several factors, such as system reliability, aging assets, equipment loading close to margin, and regulatory environment in the planning process.

7.3 Distribution Automation

Deregulation and restructuring of the electric utility business has forced utilities to turn their attention towards providing better supply reliability and quality to customers at the distribution level. Many states are requiring utilities to report their annual reliability performance in the distribution systems and some have implemented performance-based rates. Although higher reliability and quality are the goals of the utilities, they would like to accomplish this while optimizing the resources. Another goal for utilities is improvement in system efficiency by reducing system losses. DA provides options for real-time computation, communication, and control of distribution systems, and thus provides opportunities for meeting the abovementioned goals. The concept of distribution automation first came into existence in 1970s [12] and since then its evolution has been dictated by the level of sophistication of existing monitoring, control, and communication technologies, and performance and economic factors associated with the available equipment. Evolution of Supervisory Control And Data Acquisition (SCADA) systems, which have been in use for monitoring the generation and transmission systems, has also helped progress in the field of distribution automation.

Although distribution systems are a significant part of power systems and progress in computer and communication technology has made distribution automation possible [13–18], advances in distribution control technology have lagged considerably behind advances in generation and transmission control. Progress of distribution automation has been relatively slow due to reluctance of utilities in spending money on automation since many utilities have found it difficult to justify automation based purely on cost–benefit numbers. However, distribution automation provides many intangible benefits, which should be given consideration while deciding to implement distribution automation. With recent emergence of Smart Grid concept, distribution automation has come to the forefront with many utilities looking at automation of different aspects of distribution systems.

In general, functions that can be automated in distribution systems can be classified into two categories, namely, monitoring functions and control functions. Monitoring functions are those needed to record (1) meter readings at different locations in the system, (2) the system status at different locations in the system, and (3) events of abnormal conditions. The data monitored at the system level are not only useful for day to day operation but also for system planning. SCADA systems perform some of these monitoring functions. The control functions are related to operations, such as switching a capacitor, or reconfiguring feeders. In addition, system protection can also be a part of overall distribution automation schemes. Some customer related functions, such as remote load control, demand response, automated meter reading, and remote connect/disconnect may also be considered as distribution automation functions. However, automated meter reading has evolved significantly as part of AMI. Now, both DA and AMI can be considered part of the smart distribution system.

The functions mentioned above are performed in a relatively slow time frame (minutes to hours). These devices are not designed to endure frequent switching. Recently, several new devices have been developed which allow rapid control. Application of distribution-level power electronic devices such as the Static Condenser (STATCON) for distribution system control has already been demonstrated [19]. These devices are continuously controlled and respond in real-time to system changes. Coordination of a STATCON with Load-Tap-Changer (LTC) and mechanically switched capacitors reduces fluctuations in system voltage, improving the quality of service. Similarly Dynamic Voltage Restorer (DVR) can be used for critical loads to maintain the desired voltage amplitude and frequency on the load side in the event of fluctuations on the source side [20].

Implementation of DA requires careful thinking and planning. As discussed in a presentation [21], the utilities can either adopt the "top-down" approach or the "bottom-up" approach. The top-down approach is the revolutionary approach in which a large-scale fully integrated automation system is installed to automate most or all of the functions performed by various individual devices in the distribution system. The bottom-up approach is evolutionary in the sense that automation devices to perform only a particular function are installed or only a small part of the system is automated. Other functions and other parts of the system are automated gradually.

The top-down approach is expensive and requires major modifications in the utility operation, and thus, it is suitable for only a few utilities. The bottom-up strategy is more suitable for a majority of utilities. This approach allows utilities to adjust to changes at a more measured pace and to install automated systems for the most immediate needs. However, the most difficult task for a utility contemplating distribution automation is to identify the functions to be automated [14, 22]. The needs of every utility are dependent on geographic location, operating philosophy, and financial situation. Therefore, a careful screening of all the possible control functions is imperative before implementing any of them.

7.4 Distribution Automation Functions

DA functions can in general be divided into two main categories, namely customer level functions and system level functions. The customer level functions are those functions which require installation of some device with communication capability at the customer premises. These include demand response, remote meter reading, time-of-use rates, and remote connect/disconnect. The system level functions are those functions which relate to system operations. The control and communications devices for these functions are installed at different locations in the system, such as substations and feeders. These functions include fault detection and service restoration, feeder reconfiguration, voltage/var control, etc. In addition to system operation type functions, digital protection of substations and feeders can be considered part of DA in some situations.

This chapter deals only with system operation related functions. Many people prefer to subdivide these functions into two groups, namely substation related functions and feeder related functions [12]. In fact, some consider the domain of DA to include only feeder level functions; substation level functions are covered by a separate field called substation automation [23]. Although most of the focus is on feeder level functions in this chapter, such division of functions has not been considered. Each function selected may be applicable for both substation and feeders. In some situations, the functions at substation and feeder level may be performed in a coordinated fashion, for example, the switching of capacitors on the feeders may be coordinated with the switching of capacitors at the substation. A list of functions considered follows:

1. Outage management
2. Feeder reconfiguration
3. Voltage and reactive power management
4. Monitoring and control for asset management

7.4.1 Outage Management

Outage management system includes three specific tasks, which are fault location, fault isolation, and service restoration. Fault location can be based on real-time measurements from protective devices or other devices installed at strategic locations in the system. In addition to communication link with the control center, these devices have peer-to-peer communication in many cases. Newer techniques based on voltage sag and current rise are being developed. Fault location can also be based on information gathered from customers' premises from smart meters or customer calls.

Typically, the outage data gathered from customers is escalated from the lowest level of the system to the substation level to find the common point of failure. For example, if the reported data shows outage at two or more houses served by a distribution transformer, it can be concluded that the transformer had the fault. Similarly, if the reported data shows outages at several homes served by a lateral, it can be concluded that the outage is caused by a problem on the lateral. Such methods work well if there is sufficient outage data from the customers, but have difficulty in identifying precise outage locations if there are multiple outages [24]. More advanced techniques require installation of sensors on the feeders in addition to customer-ends. One such approach requires precise recording of the time of service interruption [25]. Once the location is known, the faulted section is isolated from the rest of the system with the help of remote controlled sectionalizers if the protective devices have not already isolated the faulted part. Subsequently, the switching needed to restore power to unfaulted parts of the system can be accomplished. Although in an automated system, most of the switching can be accomplished remotely, the role of humans in the fault isolation and restoration process cannot be ignored. Once identified, the faulted section has to be inspected manually by the crew to fix the problem before power can be restored to people connected to that section. Automation speeds up the process by locating outages and directing crew to precise locations instead of a general area.

Outage management takes on a different meaning for utilities in the event of a major storm, such as tornadoes, hurricanes, and ice storms. These storms can result in outage to thousands of customers. Thus, whenever such a storm hits the service territory of a utility, normal operation of the utility ceases to exist. All the resources, personnel, as well as equipment, are diverted towards repairing the system and restoring power to the customers.

Escalating the outage data from customers to locate outages is a good approach for laterals. However, for faults on primary feeders, sectionalizers with communication capabilities are used for fault location, fault isolation, and service restoration. An example system shown in Fig. 7.1 illustrates the concept. Let us consider that a fault takes place at F. Since the feeder is radial, the circuit breaker and sectionalizer A sees this fault, but sectionalizers B and C do not see it. Sectionalizer A sends a signal to Sectionalizer B informing it of the fault. The circuit breaker opens, which is followed by sectionalizers A and B opening. After this, the circuit breaker closes to restore power to the section between it and sectionalizer A. The healthy section

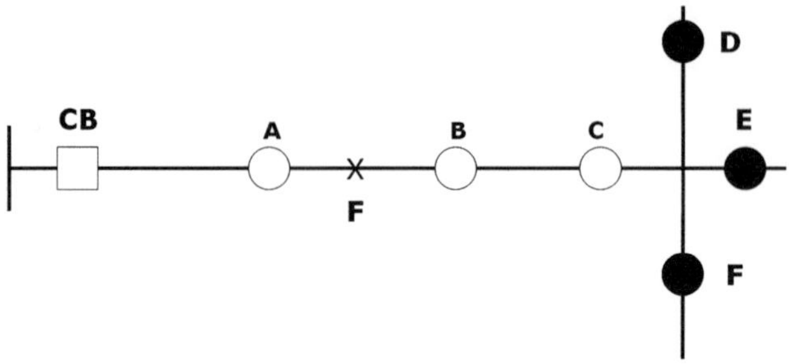

Fig. 7.1 A fully automated distribution feeder with outage location, fault isolation, and restoration capabilities. *Light circles* are normally closed sectionalizers and *dark circles* are normally open sectionalizers

between B and the normally open sectionalizers D, E, and F is still without power. Therefore, the next action is to close one of the open sectionalizers to restore power to the healthy parts of the feeder. Several rules should be followed to select the open sectionalizer to close, which are

- Look for a path from the same transformer or another transformer depending on the available spare capacity
- The feeders in the path should not be overloaded
- Voltages should be within ANSI limits
- Minimize switching operations

Additional load transfers may be required to mitigate feeder and transformer overload. This is a well researched topic with several papers available in the literature on determining the optimal path for restoration of service to the unfaulted parts of the feeders.

7.4.2 Feeder Reconfiguration

Load in a distribution system varies by hour, by day and by season. For every load level, the system has an optimal configuration of feeders. In the past, optimality had been defined in terms of minimum losses, but now service reliability has become an equally important criterion for system operation. Reconfiguration requires opening of a closed switch and closing of an open switch such that the radial structure of the system is maintained and no parts of the system are without power supply. In a manual system, the reconfiguration of system is typically done on a seasonal basis, perhaps, at the most a few times in a year or during system expansion. Since such reconfiguration may require several manual switching operations and it causes short-term interruption of service to some customers, it is not feasible to do it more frequently.

In an automated system, reconfiguration can be accomplished using the same sectionalizers which are used for fault isolation and service restoration. Since the operation of the sectionalizers is controlled remotely, system reconfiguration can be done as frequently as the dispatcher desires. From a practical point of view, however, reconfiguration poses several challenges. Firstly, reconfiguration would require closing the normally open switches, which will temporarily create a loop in the radial system. This would be followed by opening of a closed switch to bring the system back to radial configuration. If creation of a loop is not desirable, a closed switch would have to be opened first before closing the open switch, which will cause short-term interruption of service to some customers. Additionally, opening and closing of switches causes wear and tear of switches, which will reduce life of these switches. Therefore, the benefits of frequent reconfiguration must be evaluated against the cost of replacing switches.

Mathematically, the distribution system reconfiguration problem is a complex, combinatorial optimization problem involving constraints. The complexity of the problem arises from the fact that distribution network topology has to be radial and power flow constraints are nonlinear in nature. Since a typical distribution system may have hundreds of switches, an exhaustive search of all possible configurations is not a practical solution. Therefore, most of the algorithms in the literature are based on heuristic search techniques or artificial intelligence techniques.

7.4.2.1 Multi-objective Reconfiguration Problem

Several different objectives can be included in multi-objective distribution system reconfiguration problem. These objectives may include loss minimization, balancing load on transformers, balancing load on feeders to minimize loading, and deviation of voltages from nominal. Under emergencies, loss minimization is not important, but number of switching operation to complete restoration could be included as an objective. In the discussion that follows, three separate objectives, which are system loss, transformer load balance, and voltage deviation from nominal, are considered.

In multi-objective optimization, it is possible to compare two solutions by using the concept of dominance. Without loss of generality, if we assume that the optimization problem involves minimization of the objective functions, then a solution $x^* \in \Omega$ where Ω is the set of all x that satisfy all constraints, is said to be Pareto Optimal if and only if there does not exist another solution $x \in \Omega$ such that $f_i(x) \leq f_i(x^*)$ for all $i = 1, \ldots, k$ and $f_i(x) < f_i(x^*)$ for at least one i. When comparing two solutions, a solution u is said to dominate over another solution v, if and only if u is at least as good as v along with all the objectives, and furthermore, if there is at least one objective where u is better than v. In a solution space, the set of all non-dominated solutions is referred to as the Pareto set. The goal of the multi-objective optimization algorithm is to extract diverse samples from this set.

With the three objectives described above, the multi-objective distribution system reconfiguration problem can be defined as the minimization of the vector:

$$F(G) = [f_1(G)\ f_2(G)\ f_3(G)]^{\mathrm{T}} \tag{7.1}$$

where $f_1(G), f_2(G)$ and $f_3(G)$, are described below.

1. *Minimization of real power loss*:

   For a given configuration G, total real loss is defined as:

$$f_1(G) = \sum_i I_i^2 \cdot r_i \tag{7.2}$$

   where $i \in \{1, 2, \ldots, N_{\mathrm{cb}}\}$. N_{cb} is the number of connected branches in the system and r_i is the resistance of the ith branch.

2. *Transformer load balancing*:

   Loading on the substation transformers is balanced only when the load shared by each transformer in a distribution system is proportional to the capacity of that transformer. This loading is called ideal loading of the transformer and is calculated by multiplying the fractional capacity of the transformer with the sum of total loss and load (in MVA) on the network. Fractional capacity of a transformer is equal to the ratio between transformer capacity and the sum of capacities of all transformers in the system. For a given configuration G of the network, unbalance in transformer loading is measured by calculating the linear sum of absolute value of per unit deviation from the ideal loading for each transformer. Unbalance in transformer loading is defined as:

$$f_2(G) = \sum_j \mathrm{dev}_j, \tag{7.3}$$

   where $j \in \{1, 2, \ldots, N_{\mathrm{T}}\}$. N_{T} is the number of substation transformers in the system. The quantity dev_j, for the jth transformer is defined as the percentage deviation of transformer loading (LT_j) from its ideal loading, IL_j, as shown below:

$$\mathrm{dev}_j = \frac{|\mathrm{LT}_j - \mathrm{IL}_j|}{\mathrm{IL}_j}, \tag{7.4}$$

   where the ideal loading IL_j is defined as:

$$\mathrm{IL}_j = \frac{\mathrm{TC}_j}{\sum_k \mathrm{TC}_k} \cdot T_{\mathrm{LL}}, \tag{7.5}$$

where $j \in \{1, 2, \ldots, N_{\mathrm{T}}\}$ and,

$$T_{\mathrm{LL}} = \sum_p \mathrm{Load}_p + \sum_q \mathrm{Loss}_q \tag{7.6}$$

such that $p \in \{1, 2, \ldots, N_{\mathrm{b}}\}$ and $q \in \{1, 2, \ldots, N_{\mathrm{cb}}\}$. TC_j is the capacity of the jth transformer, k is the number of transformers in the system, T_{LL} is the total load plus the losses of the system, Load_p is the load on bus p, Loss_q is the loss on the qth connected branch, N_{b} is the number of buses, and N_{cb} is the number of connected branches.

3. *Minimization of voltage deviation*:

   Voltage deviation from the ideal operating value of 1 per unit is defined as:

$$f_3(G) = \max\{|1 - \min(V_i)|, |1 - \max(V_i)|\} \tag{7.7}$$

   where $i \in \{1, 2, \ldots, N_{\mathrm{b}}\}$. V_i is the voltage at the ith bus.

Illustrative Example

In this section results of applying a hybrid algorithm based on artificial immune system and ant colony optimization (AIS-ACO hybrid algorithm) [26] on a sample system are presented. The focus is not to discuss the methodology, but to discuss characteristics of distribution systems through these results. The test system [27] has a total of 86 buses with 3 substations, 8 feeders, and 96 switches. It is assumed that two transformers, each of capacity 20 MVA are located at each of the substations. Data for peak loading condition was used for simulation.

Table 7.1 presents the numerical values of the objectives for solutions having the minimum value along each objective from the Pareto set for this problem. L is the total real loss in the system (in per unit), ΔV is the maximum deviation of voltage magnitude from 1 per unit at the buses, ΔT_{b} is the sum of per unit deviation of loads

Table 7.1 Best solutions along each objective for the system

L (pu)	ΔV (pu)	ΔT_{b} (pu)	T1 (MVA)	T2 (MVA)	T3 (MVA)
Minimum loss					
0.3918	0.0303	0.544	10.995	16.808	17.493
Minimum voltage deviation					
0.3933	0.0286	0.616	10.447	16.808	18.045
0.3937	0.0286	0.543	11.000	16.808	17.493
0.3962	0.0286	0.502	11.310	15.953	18.045
0.3966	0.0286	0.429	11.862	15.953	17.493
0.4019	0.0286	0.358	12.406	15.420	17.493
Minimum unbalancing in transformer loading					
0.7966	0.1036	0.002	15.401	15.417	15.384

on transformers from ideal loading and T1, T2, T3 are the loading (in MVA) of the transformer at substations #1, #2 and #3 respectively.

It is clear from the results that not all the objectives can be minimized simultaneously. Therefore, the operators have a choice to pick one of the solutions based on their preference. Another interesting observation is that there are five solutions that give the same minimum voltage deviation. Amongst these the one that has the lowest losses has the highest transformer loading unbalance. Further, with increase in losses the transformer loading unbalance decreases. This example is a good illustration of the trade-off between different quantities in the system, which are important for system operation.

All the quantities that we have considered for reconfiguration are important for operation under normal conditions. During restoration after an outage, some of the quantities are removed from consideration and some others are relaxed. Typically, loss and unbalance in transformer loading would not be a concern during restoration and higher voltage deviation is allowed. Speed of restoration and number of switching operations are of major concern during restoration. Availability of spare capacity from the adjoining feeders and substations are taken into consideration to obtain the best restoration solutions.

7.4.3 *Voltage and Reactive Power Management*

This function is very important for utilities to provide customers with proper service voltage under different operating conditions and for minimizing losses in the system. Also, most of the customer loads are inductive, which requires utilities to place capacitors in the distribution system to provide reactive power compensation. Proper management of voltage and reactive power requires coordination between LTC at the substation transformer, switched capacitors in the distribution system, and line regulators. In this section, the role of each of these devices and their operation is discussed.

7.4.3.1 Transformer LTC Operation

LTC is a mechanical device that is built into the transformers to change the number of windings by moving the tap up or down. The LTC moves up and down based on the load on the feeder to increase the voltage under heavy load conditions and decrease the voltage under light load conditions. These are custom designed to meet the needs of individual customers. As an example, a transformer can have LTC with ±10 % control from the nominal in 5/8 % steps. This would give 16 steps on either side of the nominal. LTC also have an adjustable bandwidth around the nominal voltage and a time delay for initiating operation of taps. Mechanical tapchanger controls had limited options, but modern digital tapchanger controls provide bandwidths of 1 or 10 V in 0.1 V increments with time delay of 1–120 s in 1 s increments. LTC moves the taps if the voltage goes outside the set band and stays there for the specified time delay. Bandwidth and time delay are included to prevent frequent operation of taps. Both smaller bandwidth and smaller time delay would increase frequency of operation. LTC are designed for 500,000 operations before contact replacements [28]. So for a life of 40 years, the LTC may have 34 operations per day without a need to replace contacts. However, in actual systems, a much lower number of LTC operations takes place.

LTC operation can be controlled either based on the voltage measured on the feeder gateway at the substation or based on an estimated voltage at a specified point on the feeder. The latter is done using a technique called line-drop compensation. It uses a model to represent impedance of the feeder up to the point of control in conjunction with a voltage regulating relay that controls the taps. The voltage across the regulating relay is equal to the voltage measured at the substation minus the estimated voltage drop on the feeder. If actual measurements are not available, line impedance from the regulator to the regulated point and measured current at the regulator are used to estimate the voltage at the regulated point. Therefore, the control is only approximate without any idea of its effects on the customers at the end of the feeder. If voltage measurements can be obtained from the end of the feeder, which is possible with automation, and incorporated into control, the precise and optimal control of LTC can be achieved. This would also allow implementation of conservation voltage reduction (CVR) to reduce demand on the system.

7.4.3.2 Regulator Operation

Regulators are autotransformers and work very similar to the LTC in terms of operation and provide voltage control in ± 10 % range in 32 steps of 5/8 %. The difference is that regulators are physically separate from transformers. They can be either three-phase or single-phase and can be located in the substation or on the feeders. The ones that are designed for feeders are smaller in size and are mounted on the poles. If a system is well designed with proper selection of conductors for the feeders to match the loading, it should not need any line regulators. These are typically used in rural systems with very long feeders or in systems which have had an unexpected load growth in a specific part of the system. In the latter case line regulators are a cost effective way to address voltage related issues without having to upgrade the whole feeder. Control of regulators is done based on the voltage at the regulator or at a specific point in conjunction with line-drop compensation. Since line regulators are designed for finer control of voltage on the feeders, they could see more frequent operation than the LTC.

7.4.3.3 Capacitor Operation

Capacitors are used in distribution systems to inject reactive power, which helps in voltage control, power factor correction, and loss reduction. Capacitors can be either fixed or switched. Usually the need for capacitors increases with the increase in load. Therefore, fixed capacitors are installed based on the reactive power needs under low load conditions. Additional capacitors are switched on in steps during load increase and they are switched off in steps during load decrease. Voltage, reactive power flow, power factor, temperature, and time or a combination of these factors has been used to switch capacitors on and off.

Time-based control assumes that load will change to a certain value at a given time in the day. Although the load pattern for each day is similar, the exact load is different on each day. Therefore, time based approach works well in general, but it does not provide very precise control. Also, it does not account for holidays and weekends. The temperature-based approach assumes that the increase in load will follow the increase in temperature. This is generally true in summer when beyond a certain temperature air conditioner usage increases resulting in high load. However, not all days with similar temperature profile give the same load profile. A high temperature day in the beginning of summer could have lower air conditioner usage compared to a high temperature day in the middle of summer due to heat buildup. Also, the air conditioner load lags the temperature by 2–4 h due to thermal inertia. Similar to time-based control, temperature-based control does not account for holidays and weekends. Details on different types of controls and associated issues are available in [29].

A combination of voltage, reactive power, and power factor measurements provide a more direct way to control capacitors because an increase in load will create changes in these variables. In the past, voltage, reactive power, and power factor were measured locally near the capacitor, but with distribution automation it is possible to measure these quantities at different locations to provide better control of capacitors. With larger number of measurements from the system, more coordinated and precise control of capacitors can be obtained and coordinated with regulators and LTC. Capacitors, however, should not be switched very often because every switching operation generates a spike of current. Therefore, frequent operation can lead to failure of the switch controlling it or the failure of the capacitor itself. Capacitor switching has also shown to create power quality issues including harmonics. Devices such as STATCOM (static compensator) provide much better and precise control, but they are significantly more expensive. Therefore, they are used only in very special situations.

7.4.4 Monitoring and Control for Asset Management

The purpose of distribution system monitoring is very similar to SCADA in the traditional sense. Monitoring is necessary to acquire data for many of the distribution functions. Some of these functions require real-time data from the system to make control decisions. Real-time data is also useful in providing information to operators on abnormal system conditions in the form of alarms. In addition to the real-time data, system data can be gathered and archived for later use. Such data can then be used for forecasting and planning purposes. Monitoring is also useful for management of major assets, such as substation transformers and circuit breakers, in distribution systems as described in the following sections.

7.4.4.1 Transformer Life Extension

The substation transformers normally operate at loads lower than their capacity. However, during emergencies, such as failure of another transformer, they can be operated at loads higher than the rated capacity. But overloading can be done only for a limited time without substantially affecting the life of the transformer. The higher the overloading, the lower is the time allowed for overloading. In a manual process, the dispatcher has to rely on trial-and-error methods to get the proper level of loading. The dispatcher would close the switch to supply additional load with an expectation that the total load is less than a certain value. But if the load after switching is higher than expected, he would have to open the switch, drop a few feeders, and then close the switch again. The process would have to be repeated until the load is at a desired level. The switching on and off of load can stress the transformer significantly and thus reduce its total life. Using an automated procedure, this task can be performed without stressing the transformer.

Automation of this function requires equipment for monitoring the transformer including oil and winding temperatures. Equipment for monitoring the health of the transformer based on dissolved gas analysis is also available. Data and measurements from the feeders connected to the transformer are needed too. The oil and winding temperatures determine the level of overloading possible under the given loading conditions. Then the feeders can be selected such that there is a balance between the desired loading and the loads of the feeders. Thus, overloading of the transformers can be controlled precisely without too many unwanted switching operations. Hence, stress on the transformers can be avoided and life extension of transformers can be achieved. Such controlled loading is very useful during restoration following extended outages to mitigate effects of cold load pickup [30].

7.4.4.2 Recloser/Circuit Breaker Monitoring and Control

Typically no remote monitoring and control is available on the breakers and reclosers. The settings of the relay and recloser timings can be changed only by going to the location of the equipment. In case of pole mounted reclosers, it is extremely time consuming to change settings. Further, since no monitoring is available, the recloser and breaker contacts are refurbished at fixed intervals whether it is necessary or not. This maintenance frequency is usually based on the duty level the recloser or breaker is expected to perform. Generally, the maintenance interval is estimated conservatively (i.e., refurbishments are made, on the average, sooner than is necessary). Hence, in many cases the contacts are serviced before it is necessary.

The advantages of automating this function are many. In an automated scenario, firstly the relay settings and recloser timings can be set remotely. This will allow for better control of the system whenever the system configuration changes. Moreover, the labor needed to reset the relay and recloser timings can be saved because these settings can be done remotely instead of going to the location. Secondly, monitoring of the energy interrupted by the recloser and breaker can provide a precise estimate of health of the contacts. Using this information refurbishing of the contacts can be scheduled whenever necessary. Hence, too early or too late servicing of the recloser and breakers contacts can be avoided.

7.5 Cost–Benefit of Distribution Automation

Determining costs and benefits associated with different distribution automation functions is very important for making decisions regarding implementation of these functions [31]. Cost for automation includes cost for communication infrastructure, hardware, monitoring and control equipment, installation, software, training, and operation and maintenance. Benefits of automation are obtained in three broad categories, which are higher efficiency, higher reliability, and higher quality. Higher efficiency includes lower system losses, lower peak power requirement, and reduced manpower needs. Higher reliability includes lesser and shorter sustained power interruptions to the customers, failure avoidance of equipment, and balanced loading of equipment. Higher power quality includes lesser momentary interruptions, lesser voltage sags and swells, lesser incidences of voltage out of range at the customer-ends and lowered THD_v (total harmonic distortion in voltage). Benefits are usually accrued annually after implementation of distribution automation, whereas cost has two components, which are capital cost and annual operating and maintenance expenses. Some benefits and costs may increase at a certain rate annually.

The first step in determining benefits is to map DA application to operational benefits. Northcote-Green provides details of such mapping in his book on distribution automation [32]. Based on a similar approach, mapping of functions discussed in the previous section is shown in Table 7.2. The next step is to map operational benefits to monetary benefits. Outcomes of automation must be compared with the status quo or a traditional improvement scheme to obtain benefits or costs. If actual values of some of the variables to determine monetary benefits or costs are not available, the planners have to use an estimated value for these parameters. In case of uncertainty, different values of some variables can be tried to get a range of answers. Also, sensitivity analysis of the outcomes with respect to the input variables gives an indication of the importance of certain variables on the cost–benefit analysis. Thus, planners should focus on those variables to which the cost–benefit analysis is more sensitive.

To illustrate the computation of monetary benefits of different operation benefits, let us consider a few examples of operation benefits of outage management.

Table 7.2 Mapping the benefits of distribution automation functions

	Higher energy sales	Lower SAIFI	Lower SAIDI	Life extension of Equipment	Lower losses	Reduced peak demand	Reduced manpower	Lesser low-voltage complaints
Outage management	X	X	X	X			X	
Feeder reconfiguration			X	X	X	X		X
Voltage/VAr management					X	X	X	X
Monitoring and control for asset management		X	X	X			X	

Higher Energy Sales:

$$\text{Annual benefit} = (\$/\text{kWh})(\text{average customer demand})(\%\text{ of system lost}/100)$$
$$(\text{number of customers})(\text{line failures per year per mile})$$
$$(\text{circuit miles})(\text{average decrease in outage hours per event})$$

Reduced Labor for Fault Location:

$$\text{Annual benefit} = (\text{line failures per year per mile})(\text{circuit miles})\begin{bmatrix}(\text{line crew man hours per fault})\\(\text{line crew }\$/\text{man hour})\\+(\text{operator man hours per fault})\\(\text{operator }\$/\text{man hour})\end{bmatrix}$$

O&M of Switches and Controllers:

$$\text{Annual benefit or cost} = \begin{bmatrix}(\text{manual switches per feeder})\\(\text{failure rate of manual switches})\\(\text{repair cost per manual switch})\\-(\text{automated switches per feeder})\\(\text{failure rate of automated switches})\\(\text{repair cost per automated switch})\end{bmatrix}(\text{number of feeders})$$

If this number turns out to be positive, it will be a benefit. But if it turns out to be negative, it will be a cost.

Lesser Low-Voltage Complaints:

$$\text{Annual benefit} = \left(X - X'\right)(\text{Cost per complaint})(\text{Number of customers})/1{,}000$$

where X = Complaints without automation per thousand customers per year and X′ = Complaints with automation per thousand customers per year.

7.6 Conclusion and Future Directions

Distribution systems have evolved significantly over the years. From a passive radial system they have become active system with substantial embedded generation and automation. Operation of the future distribution systems will become more complex. As a result of this, smart distribution automation systems will be needed to manage them. These systems will require faster decisions and thus real-time analysis utilizing large amounts of data. Applications expected to be integrated in the smart distribution systems include:

- Optimal Volt/VAr optimization
- Online power flow and short circuit analysis
- Advanced and adaptive protection
- Contingency analysis

- Advanced fault detection and location
- Advanced fault isolation and service restoration
- Automated vehicle management system
- Dynamic derating of power equipment due to harmonic content in the load
- Distribution operator training simulator
- System operation with large penetration of customer-owned renewable generation
- Distribution system operation as a stand-alone microgrid
- Real-time pricing and demand response applications

Significant new research and development in smart distribution systems is needed for integration of the abovementioned applications.

Acknowledgment The author would like to thank many colleagues and graduate students with whom he has worked for over 30 years. Collaboration with them has resulted in some of the material presented in this chapter. Specifically, the author would like to thank Ashish Ahuja and Dr. Sanjoy Das for material related to multi-objective reconfiguration, and Dr. J. Kenneth Shultis for formulation of equations for cost–benefit of distribution automation functions.

References

1. Willis HL (1997) Power distribution planning reference book. Marcel Dekker, Inc., New York, NY
2. The Smart Grid: an introduction. U. S. Department of Energy Publication, 2008
3. IEEE Power & Energy Magazine. Special issue on smart grid. March/April 2009
4. Brown HE, Suryanarayanan S, Heydt GT (2010) Some characteristics of emerging distribution systems under the Smart Grid Initiative. Electr J Elsevier 23(5):64–75
5. Malekpour AR, Pahwa A, Das S (2013) Inverter-based var control in low voltage distribution systems with rooftop solar PV. 45th North American Power Symposium, Manhattan, KS
6. Abbey C, Cornforth D, Hatziargyriou N, Hirose K, Platt G, Kwasinski A, Kyriakides E, Reyes L, Suryanarayanan S (2014) Powering through the storm: microgrids operation for more efficient disaster recovery. IEEE Power Energy Mag 12(3):67–76
7. Brown RE (2002) Electric power distribution reliability. Marcel Dekker, New York, NY
8. Heydt GT (1991) Electric power quality. Stars in a Circle Publications, West Lafayette, IN
9. Faruqui A (2007) Breaking out of the bubble. Public Utilities Fortnightly, March
10. IEEE Application Guide for IEEE Std 1547-2 (2008) IEEE Standard for interconnecting distributed resources with electric power systems. IEEE, New York
11. Pahwa A, DeLoach SA, Das S, Natarajan B, Ou X, Andresen D, Schulz NN, Singh G (2013) Holonic multi-agent control of intelligent power distribution systems. IEEE PES General Meeting, Vancouver, BC
12. Clinard K, Redmon J (eds) (1998) Distribution management tutorial. IEEE PES Winter Meeting, Tampa, FL
13. Pahwa A, Shultis JK (1992) Assessment of the present status of distribution automation. Report No. 238, Engineering Experiment Station, Kansas State University, Manhattan, KS, March 1992
14. Bassett D, Clinard K, Grainger J, Purucker S, Ward D. Tutorial course: distribution automation. IEEE Publication 88EH0280-8-PWR
15. Moore T (1984) Automating the distribution network. EPRI J September:22–28

16. Moore T, Bunch JB (1984) Guidelines for evaluating distribution automation. EPRI Report EL-3728, November
17. Kendrew T (1990) Automated distribution. EPRI J January/February:46–48
18. Bunch JB (1984) Guidelines for evaluating distribution automation. EPRI Report EL-3728, November
19. Paserba JS, Miller NW, Naumann ST, Lauby MG, Sener FP (1993) Coordination of a distribution level continuously controlled compensation device with existing substation equipment for long term var management. Paper no. 93 SM 437-4 PWRD, IEEE PES summer meeting, Vancouver, Canada, July 1993
20. Suryanarayanan S, Heydt GT, Ayyanar R, Blevins JD, Anderson SW (2008) Simulation based considerations in placement of capacitors near a dynamic voltage re-storer. Simul Model Pract Theory 16(9):1430–1437
21. Undren EA, Benckenstein JR (1990) Protective relaying in integrated distribution substation control systems, Presentation for panel session on integration of demand-side management and distribution automation. IEEE Power Engineering Society Winter Meeting, Atlanta, Georgia
22. Davis EH, Grusky ST, Sioshansi FP (1989) Automating the distribution system: an intermediary view for electric utilities. Public Utilities Fortnightly 19:22–27
23. Block D (1996) Utility automation technology, Electric power industry outlook and atlas 1997 to 2001. PennWell Books, Tulsa, OK
24. Laverty E, Schulz NN (1999) An improved algorithm to aid in post-heat storm restoration. IEEE Trans Power Syst 14(2):446–451
25. Rodrigo PD, Pahwa A, Boyer JE (1996) Location of outages in distribution systems based on hypotheses testing. IEEE Trans Power Deliv January:546–551
26. Ahuja A, Das S, Pahwa A (2007) An AIS-ACO hybrid approach for multi-objective distribution system reconfiguration. IEEE Trans Power Syst 22(3):1101–1111
27. Schmidt HP, Kagan N (2005) Fast reconfiguration of distribution systems considering loss minimization. IEEE Trans Power Syst 20(3):1311–1319
28. IEEE standard requirements for tap changers. IEEE Std C57.131™-2012, IEEE, New York, 2012
29. Coughlan BW, Lubkeman DL, Sutton J (1990) Improved control of capacitor bank switching to minimize distribution system losses. The proceedings of the twenty-second annual North American Power Symposium, Oct 90. pp. 336–345
30. Ucak C, Pahwa A (1994) An analytical approach for step-by-step restoration of distribution systems following extended outages. IEEE Trans Power Deliv July:1717–1723
31. Shultis JK, Pahwa A (1992) Economic models for cost/benefit analysis of eight distribution automation functions. Report no. 234, Engineering Experiment Station, Kansas State University, Manhattan, KS, June
32. Northcote-Green J, Wilson R (2007) Automation and control of electrical power distribution systems. CRC Taylor & Francis, Boca Raton, FL

Chapter 8
Legacy of Professor G.T. Heydt to Power Engineering Research, Education, and Outreach

S.S. (Mani) Venkata

8.1 Introduction

I have known Prof. Jerry Heydt since the early 1970s as a contemporary colleague. I consider him not only a close and congenial friend but, more importantly, a mentor and inspiring colleague. I have learned a lot from my close association with him. Knowing him is a great asset in my professional as well as in my personal life. We have collaborated in all the three important domains of education, research, and outreach for over 40 years. In this chapter, I endeavor to cover the rich experiences of my close association with him in all these three areas that are the very essence of a successful academic professional. Though we now have both entered the stage in life that most would consider close to retirement, his Mantra has been that "Retirement is for quitters." I whole-heartedly agree with that sentiment. Prof. Heydt not only keeps himself busy professionally but also continues to work very hard. I provide the details of my rich experiences with him in the following domains.

8.2 Power and Energy Research

In the power and energy research domain, we collaborated chronologically in the following topics: six-phase power transmission, power quality, modeling of highly nonlinear loads, and advanced generation technologies.

S.S. (Mani) Venkata (✉)
Alstom Grid Inc., University of Washington, Seattle, WA, USA

Iowa State University, Ames, IA, USA
e-mail: mani.venkata@alstom.com

E. Kyriakides et al. (eds.), *Electric Power Engineering Research and Education*,
Power Electronics and Power Systems, DOI 10.1007/978-3-319-17190-6_8

8.2.1 Multi-phase Power Transmission

I started my academic career in 1971 and I continued research on my dissertation topic, *Sensitivity of power system transient stability using Popov's technique*. I realized quickly that this was not the field where I could seek and obtain external funding to continue my work. I quickly changed my direction of work based on the advice of Tom Lee at GE. In 1974, I was fortunate to obtain an initiation grant from the National Science Foundation to work on the topic of multi-phase power transmission with the focus on six-phase systems. This was a relatively unknown area and least explored. With the strong support of Bill Guyker from Allegheny Power, we obtained very interesting results. This work was further extended by Allegheny Power for another 3 years to explore the possibility of converting existing double-circuit 138-kV transmission lines to 138-kV six-phase transmission lines with minimal changes. In this way, the capacity of the corridor could be increased by about 1.732, thus alleviating the need to get additional right-of-way. Based on this work, Bhatt, Guyker, Booth, Guyker's colleague, and I published our first paper: *Six-phase (multi-phase) Power transmission lines: fault analysis* [1]. During this work, I had been sharing the results with Heydt. In this paper, our group derived the *Six-Phase Symmetrical Components Transformation and its application to fault analysis*. Heydt's creativity was revealed in the form of the detailed discussion reported below. He provided the systematic derivation of Six-Phase Clarke's Transformation for six-phase and eight-phase systems. The procedure can, in fact, be extended to any higher-order phase as Charles Fortescue [2], Edith Clarke [3] and other early researchers implied. But Heydt provided the detailed derivations for higher order systems for various real transformations in the *Discussion* reproduced below. (Only the relevant portions of the entire discussion and authors' response are included here).

G.T. Heydt and R. C. Burchett (Purdue University, West La Fayette, Indiana): Bhatt, Venkata, and Guyker have prepared a thoroughly interesting and thought provoking paper on 6ϕ fault analysis. There are so many interesting extensions of 3ϕ calculation techniques to investigate—and the authors have covered symmetrical component fault analysis. In this discussion, we would like to raise questions associated with a real transformation rather than a complex transformation. Since the symmetrical component transformation is complex, the assumption of sinusoidal steady state analysis is implicit in every phasor calculation. Motivation for an all real transformation is the potential application to transient analysis. In 3ϕ terminology, one well known real transformation is Clarke's transformation. Of course, all diagonalization procedures on the Z matrix yield the same sequence impedances since the eigenvalues of any square matrix are unique [1,2].

In 3ϕ circuits, the eigenvalues of Z,

$$Z = \begin{bmatrix} z_{ss} & z_{sm} & z_{sm} \\ z_{sm} & z_{ss} & z_{sm} \\ z_{sm} & z_{sm} & z_{ss} \end{bmatrix}$$

are z_0, z_1, z_2,

$$z_0 = z_{ss} + 2z_{sm}$$

$$z_1 = z_{ss} - z_{sm}$$

$$z_2 = z_{ss} - z_{sm}$$

For n phase circuits ($n\phi$), when Z is of a similar form, the eigenvalues are

$$z_0 = z_{ss} + (n-1)z_{sm}$$

$$z_i = z_{ss} - z_{sm}, \; i = 1,2,3,\ldots,n-1 \qquad (1)$$

as can be verified by substitution into

$$\det(Z - \lambda I) = 0$$

where λ is the eigenvalue, I is the identity matrix, and det (.) denotes determinant. For 3ϕ circuits, the well known transformation is Clarke's transformation, T_{c3},

$$T_{c3} = \begin{bmatrix} 1/\sqrt{3} & \sqrt{2}/\sqrt{3} & 0 \\ 1/\sqrt{3} & -1/\sqrt{6} & 1/\sqrt{2} \\ 1/\sqrt{3} & -1/\sqrt{6} & -1/\sqrt{2} \end{bmatrix}$$

where the corresponding eigenvalues (i.e., sequence impedances) are z_0, z_1, z_2 in Eq. (1). In order to extend this transformation to 6ϕ and $n\phi$ cases, note that each column of T_{cn} (the $n\phi$ real transformation) is an eigenvector corresponding to some eigenvalue. If e_i is column i of T_{cn}, then the corresponding eigenvalue is z_i. For i = 1

$$z_o = z_{ss} + (n-1)\, z_{sm}$$

and the eigenvector equation [2,3] is

$$(Z - \lambda_i I)\, e_i = 0$$

$$(Z - (z_{ss} + (n-1)\, z_{sm} I))e_1 = 0. \qquad (2)$$

Substitute the form of Z into Eq. (2),

$$\begin{bmatrix} -(n-1)z_{sm} & z_{sm} \cdots & z_{sm} \\ z_{sm} & -(n-1)z_{sm} & z_{sm} \\ & & \\ z_{sm} & z_{sm} & -(n-1)z_{sm} \end{bmatrix} e_1 = 0 \qquad (3)$$

select all elements of e_1 equal and let $\|e_1\| = 1.0$. Then,

$$\begin{bmatrix} 1/\sqrt{n} \\ 1/\sqrt{n} \\ \cdots \\ 1/\sqrt{n} \end{bmatrix} = e_1 \qquad (4)$$

All other eigenvalues are repeated and are equal to $z_{ss}-z_{sm}$. The eigenvector equation is

$$(Z-(z_{ss}-z_{sm})\ I)e_i = 0 \qquad i=2,3,\ \ldots\ ,n.$$

Substitute for Z,

$$\begin{bmatrix} z_{sm} & z_{sm} & \cdots & z_{sm} \\ z_{sm} & z_{sm} & \cdots & z_{sm} \\ \cdots & & & \\ z_{sm} & z_{sm} & \cdots & z_{sm} \end{bmatrix} e_i = 0$$

Therefore, the sum of the elements of e_i is zero. Also, e_i is orthogonal to e_1 and all other eigenvectors. A natural choice of values for e_i is arbitrary values in the top i rows, and all equal values in the bottom n-i rows. Thus e_2 is of the form $[e_{12}\ \ e_{22}\ \ e_{22}\ \ldots\ e_{22}]^T$, e_3 is $[e_{13}e_{23}e_{33}e_{33}\ \ldots\ e_{33}]^T$ etc. Using the fact that

$$\| e_i \| = 1$$

and that

$$e_i \cdot e_j = 0 \qquad i=j$$

results in the following solution for e_j, (i=2,3, . . ,n),

$$e_i = \begin{bmatrix} 0 \\ 0 \\ 0 \\ \dfrac{(n-i+1)}{\sqrt{q}} \\ \dfrac{-1}{\sqrt{q}} \\ \dfrac{-1}{\sqrt{q}} \\ \cdots \\ \dfrac{-1}{\sqrt{q}} \end{bmatrix} \begin{matrix} \\ \\ \leftarrow \text{row } i-2 \\ \leftarrow \text{row } i-1 \\ \leftarrow \text{row } i \\ \\ \\ \leftarrow \text{row } n \end{matrix} \qquad (5)$$

$$q=n^2-2ni+i^2+3n-3i+2$$

This column i of T_{cn}.

Thus the 6ϕ Clarke transformation is T_{c6},

$$T_{c6} = \begin{bmatrix} 1/\sqrt{6} & 5/\sqrt{30} & 0 & 0 & 0 & 0 \\ 1/\sqrt{6} & -1/\sqrt{30} & 4/\sqrt{20} & 0 & 0 & 0 \\ 1/\sqrt{6} & -1/\sqrt{30} & -1/\sqrt{20} & 3/\sqrt{12} & 0 & 0 \\ 1/\sqrt{6} & -1/\sqrt{30} & -1/\sqrt{20} & -1/\sqrt{12} & 2/\sqrt{6} & 0 \\ 1/\sqrt{6} & -1/\sqrt{30} & -1/\sqrt{20} & -1/\sqrt{12} & -1/\sqrt{6} & 1/\sqrt{2} \\ 1/\sqrt{6} & -1/\sqrt{30} & -1/\sqrt{20} & -1/\sqrt{12} & -1/\sqrt{6} & -1/\sqrt{2} \end{bmatrix}$$

To find column 1 in the nϕ Clarke's transformation, use Eq. (4). To find columns 2,3, . . . , n, use Eq. (5). For example, the 8ϕ transformation is

$$T_{c8} = \begin{bmatrix} 1/\sqrt{8} & 7/\sqrt{56} & 0 & 0 & 0 & 0 & 0 & 0 \\ 1/\sqrt{8} & -1/\sqrt{56} & 6/\sqrt{42} & 0 & 0 & 0 & 0 & 0 \\ 1/\sqrt{8} & -1/\sqrt{56} & -1/\sqrt{42} & 5/\sqrt{30} & 0 & 0 & 0 & 0 \\ 1/\sqrt{8} & -1/\sqrt{56} & -1/\sqrt{42} & -1/\sqrt{30} & 4/\sqrt{20} & 0 & 0 & 0 \\ 1/\sqrt{8} & -1/\sqrt{56} & -1/\sqrt{42} & -1/\sqrt{30} & -1/\sqrt{20} & 3/\sqrt{12} & 0 & 0 \\ 1/\sqrt{8} & -1/\sqrt{56} & -1/\sqrt{42} & -1/\sqrt{30} & -1/\sqrt{20} & -1/\sqrt{12} & 2/\sqrt{6} & 0 \\ 1/\sqrt{8} & -1/\sqrt{56} & -1/\sqrt{42} & -1/\sqrt{30} & -1/\sqrt{20} & -1/\sqrt{12} & -1/\sqrt{6} & 1/\sqrt{2} \\ 1/\sqrt{8} & -1/\sqrt{56} & -1/\sqrt{42} & -1/\sqrt{30} & -1/\sqrt{20} & -1/\sqrt{12} & -1/\sqrt{6} & -1/\sqrt{2} \end{bmatrix}$$

We would appreciate the authors' comments on this real transformation and potential uses in transient analysis of nϕ systems. Are the authors working on transient analysis for 6ϕ systems?

We hope that Bhatt, Venkata, and Guyker continue to report on their results in this area.

REFERENCES

[1] M. S. Chen, W. E. Dillon, "Power System Modeling", Proc. IEEE, July 1974, pp. 901-915.

[2] C. T. Chen, "Introduction to Linear System Theory", Holt, Rinehart, Winston, New York, 1970.

[3] R. Roy, C. DeRusso, R. Close, "State Variables for Engineers", John Wiley and Sons, 1965.

Navin B. Bhatt, Subrahmanyam S. Venkata, William C. Guyker, and William H. Booth: The authors are extremely thankful to the discussors for their valuable contributions.

> The purpose of conducting fault analysis is to generate data required by relay engineers in the design of protective scheme for transmission lines. In this respect, the approach being taken for six-phase lines will be similar to three-phase lines.
>
> Both Drs. Heydt and Yu raised questions on transformations. The authors are working on fault analysis and have decided to use symmetrical component transformations until it leads to unmanageable computational complexity since it has great practical and physical significance to engineers considering its wide usage. It has been recently realized by Dr. E. K. Stanek of West Virginia University and the authors that for certain types of faults real transformations like the one suggested by Dr. Heydt and Mr. Burchett in their discussion offer more advantages than the symmetrical components. It is possible that "phase coordinates" concept may prove to be more valuable for system fault analysis than the various transformations mentioned above. It is too early to make clear judgements at this time with regard to choice on the transformations. Further study is required in this regard. It is also too early to say whether simultaneous fault analysis of six-phase lines would need to be performed for suitable protective scheme design.

From the mid-1970s, Heydt has shown keen interest in mathematical transformations for higher-phase order (HPO) systems. He has recently encouraged one of his doctoral students at Arizona State University (ASU) to pursue research work in this area. My interest also continues in the area but my focus is more on extending this concept to multi-phase power distribution systems. I strongly believe that it could be extended to both primary and secondary voltage levels both from right-of-way reduction as well as overall cost reduction. The ideas are still in embryonic stage and I am looking for an opportunity to spend some time with Heydt to pursue our interests in this field. It is rather unfortunate that the concept of HPO has not gained much acceptance from utilities for more than four decades since it was introduced.

8.2.2 Power Quality

This phrase is synonymous with Heydt and his name and fame is in the area. He has published a classic text book in this field [4]. This is adopted by many universities as a senior/graduate level text book. It is also used by many researchers and engineers as a reference book. As he states in the *Introduction*, "the emphasis of the text book is on the description, analysis, modeling and the solution for difficulties relating to distortion of wave shape in alternating current power systems." For those who are not familiar with the contents of this book, it has 10 chapters. In Chap. 1, Heydt starts with the question: *Electric Power Quality: What is it?* He then proceeds systematically to cover *Measures and Standards for Power Quality* in Chap. 2 and *Measurements* in Chap. 3. In Chap. 4 he focuses on *Modelling of Networks and Components Under Nonsinusoidal Conditions*. It is followed by the topic *Loads Which May Cause Power Quality Problems* in Chap. 5. This is logically followed by *Analysis Methods* in Chap. 6. Chapter 7 covers a very important problem, *Harmonics in Power Systems*. In this chapter, he covers his original work on *Harmonic Power Studies* in detail. This work paved the way for his legendary work on power quality. Chapter 8 deals with *State Estimation Applied to Power Quality Issues*. Perhaps the least understood matter on *Impulses, Radio Frequency Signals and Flicker* is the topic of Chap. 9. The book concludes with the topic of *Power Quality Improvement* in Chap. 10. It is a must read for all power systems engineers.

He wrote this book while he was on a sabbatical from Purdue University at Avaruva, Rarotonga, Cook Islands in South Pacific. This is a relatively unknown but a desirable location to undertake such an involved project. He went there at the invitation of the Government of Cook Islands to provide technical assistance to them. This is a clear example of his professional reputation in power systems all around the world.

In 1990, Heydt and I proposed to the National Science Foundation to hold a workshop in this field to attract and encourage young researchers and educators to work in power quality. This realization came to us when we recognized that power electronics would play a dominant role in power system control. NSF provided funding for about 20 professors to attend this workshop in Grand Canyon, Arizona [5]. (As a side it is interesting to note, that though Heydt probably did not know at that time, he would move from Purdue University to Arizona State University in 1994 where he now holds the prestigious title of Regents' Professor.) This workshop was conducted in an unusual manner due to the creative thinking of Heydt. Instead of conducting the workshop in a classical, class room environment, it was scheduled as a serious chat session during trekking for a half a day followed by a fire side chat discussion after supper. The participants enjoyed the unusual settings for the workshop, and the workshop turned out to be a very successful event. The final report [6] was submitted and later it was published as a paper in the IEEE Transactions [7] with Professors Domijan and Meliopoulos. These two documents are referred to even today by many educators with a keen interest in power quality. With the advent of the *Smart Grid*, Heydt's original work takes even a deeper meaning because the power quality issues are gaining more prominence in future distribution system planning and operation. Personally, I am using his contributions in my current work on the real-time operations of emerging distribution systems at Alstom Grid. The infusion of many renewable generating sources such as solar, battery storage, fuel cells will need creative solutions to ever increasing power quality problems. For example, some utilities are monitoring the power quality of their substations up to the 15th harmonic in real-time.

Subsequent to the workshop, our joint interests turned to offering successful short courses to Tacoma Public Utilities, Central Intelligence Agency, and Puget Sound Energy with Prof. Meliopoulos [8] during the mid-1990s. Heydt continues his work in this area even today by offering many courses and supervising several graduate students at ASU. His book is a great resource for training courses on distribution automation that I offer to engineers throughout the world.

8.2.3 Modeling Highly Nonlinear Loads

During 1997, Electric Power Research Institute (EPRI) provided our joint team funding to develop an EMTP module to model highly nonlinear loads, such as arc furnaces, which are sources of voltage flicker problems [9]. We successfully delivered the module to EPRI [10]. But our curiosity of modeling these devices gave us a hunch that they are nonlinear and follow a chaotic process. Realizing that we need strong mathematicians to provide the theoretical foundation for modeling, we recruited Kostelich from ASU and Athreya from Iowa State University (ISU) to work with the project team. We received funding from NSF [11] to investigate this difficult problem. In many ways Heydt's expertise in power quality culminated in securing funding for this 2-year effort. We earnestly started conducting the research in both sites on a cooperative basis. It is in fact Heydt's idea to recruit the two mathematics professors for this research effort, which is another example of his creativity. The two teams collaborated very successfully by having weekly conference calls and site visits. In one of the visits to ASU, Heydt again used his creativity and took all of us on a picnic. Though we had a lot of fun that Sunday, we ended up pursuing several ideas for this research effort. He definitely has the gift for combining fun with serious work.

After 3 years of productive work on this topic, we successfully published three papers [12–14]. We also ended up offering an on-line graduate course on this topic using the latest digital delivery technology available at that time. More than 15 students took this course from both campuses. We were glad that NSF encouraged us to disseminate our results by these means to a larger community.

To highlight his work, the abstract of the paper [12] is extracted here for the benefits of those who are interested in gaining a deeper understanding of this phenomenon. In fact, I would encourage them to go through the entire paper.

"Typically, the modeling of highly varying, nonlinear loads, such as electric arc furnaces, has involved stochastic techniques. This paper presents the use of chaotic dynamics to describe the operation of nonlinear loads. Included is a discussion of the Lyapunov exponents, a measure of chaotic behavior. The alternate approach is applied to electric arc furnaces. A tuning mode is described to develop the parameters of a chaotic model. This model is trained to have time and frequency responses that are tuned to match the current from the arc furnace under study. The simulated data are compared to actual arc furnace data to validate the model. This model is used to assess the impact of various highly varying nonlinear loads that exhibit chaos in power systems."

8.2.4 Advanced Generation Technologies

In 1991, as Guest Editors, Heydt and I proposed to develop a special issue on *Advanced Generation Technologies* to the Proceedings of the IEEE Editorial Board. This was the time new generation technologies started gaining interest in the power systems community. Sensing the need to bring focus on all these technologies in one place, we proposed this issue and it was quickly accepted by the Board. We then invited the leading researchers to contribute a comprehensive and survey type of paper for the benefit of all engineers and scientists within the IEEE community. It was published in March 1993 [15]. Fifteen articles covering motivation and selection process in electric utilities, ocean thermal energy conversion (OTEC), economic assessment, tidal power, superconducting generators, battery energy storage technologies, geothermal, photovoltaic, high-efficiency electrochemical battery, wind energy systems, and magnetic fusion energy. All the contributors were way ahead of their time in foreseeing the potential for these newer technologies. Many of them are renewable in nature and they have taken deeper implications in the context of the *smart grid.*

Heydt wrote the article on ocean thermal energy conversion (OTEC), a technology about which most of the people had no idea at that time. His paper [16] is titled *An assessment of ocean thermal energy conversion as an advanced electric generation methodology.* Here is the abstract of the paper: "The history of ocean thermal energy conversion (OTEC), a process that employs the temperature difference between surface and deep ocean water to alternately evaporate and condense a working fluid, is reviewed. In the open-cycle OTEC configuration, the working fluid is seawater. In the closed-cycle configuration, a working fluid, such as propane is used. OTEC is assessed for its practical merits for electric power generation. Because rather large amounts of seawater and working fluid are required, the energy requirements for pumping them may be greater than the energy recovered from the OTEC engine itself. The concept of net power production is discussed. The components of a typical OTEC plant are described with emphasis on the evaporator heat exchanger. Operation of an OTEC electric generating station is discussed, including transient operation. Recent experiments and efforts at the National Energy Laboratory-Hawaii (NELH) are summarized. Remarks are made on bottlenecks and the future of OTEC as an advanced electric generation methodology."

The lucid and clear presentation has convinced me that Heydt can take a very difficult topic and make it understandable to those who have minimal knowledge of the subject. Personally, it was indeed an enriching experience for me to be involved in this project with him as a co-Guest Editor and developing the *Scanning the issue* article.

8.2.5 Microgrid Automation

As the future of distribution engineering is one of Heydt's active research pursuits, I would like to describe my current work on *microgrid automation* since it falls within this general area. Here again I am hoping to seek his advice as the work proceeds. The concept of *Microgrids* has gained a lot of attention recently in order to improve the overall smart distribution grid's reliability and resilience. As the name *microgrids* implies, they are formed as sub-systems within a distribution system. Their capacity could vary from a few kva to tens of Mva. For example, a typical substation could be carved into three or four microgrids, depending on the location and size of distributed generation and loads. The key requirement is to make sure that the power balance for each microgrid is achievable within the voltage and frequency standards. They need to be identified a priori to enable optimal planning and operation of the entire distribution system. The key principle of microgrid automation is the integration of sensing, control and protection. Each item is addressed below.

Sensing: The first requirement of selecting a microgrid is to make sure that it has adequate monitoring devices, either classical ones or modern ones, such as optical or micro sensors with communication capabilities.

Control: The goal is to define a fully comprehensive scheme consisting of end to end functions for operational planning and real-time operations. The functions will include microgrid islanding, synchronization and reconnection, voltage range, frequency range, power quality management, portfolio dispatch management, and enhanced system reliability and resilience coordination besides sensing and protection.

Protection: The protection scheme should provide a comprehensive approach to protecting microgrids with islanding capability. Our approach is to design a distributed differential power protection scheme requiring an optimal protection device placement algorithm, and a fault detection and location algorithm. This novel microgrid protection method can effectively detect and locate all fault types. The measurements from the protective devices are then used to locate the fault location with high accuracy. Such a fault location algorithm with a phasor measurement unit (PMU) placed at a feeder head [17] will reduce outage time by increasing restoration speed. Since devices having the characteristics needed to deploy this scheme are currently available, it is a practical and viable solution that can be implemented on any microgrid system.

8.3 Power Engineering Education

8.3.1 Promotion

As I have mentioned at the beginning, Heydt's contributions to power engineering education are priceless and too numerous to mention here. I had the opportunity to interact with him as member of the Administrative Committee of the Power Engineering Education Committee (PEEC) since we started our academic careers in the early 1970s. He has made unique and significant contributions to PEEC. He proposed the formation of the Research Subcommittee to bring all the developments happening in the power and energy field for the benefit of the members. This subcommittee continues to provide funding opportunities to all the faculty members of our field, particularly to those entering the academic profession. I should emphasize that it has been very difficult to seek funding from federal, state, utilities and other industries since opportunities were very limited until the dawn of the *Smart Grid*. But the leadership of the PEEC provides avenues for all faculty members interested in research and education. Heydt deserves the major share of credit for this initiative and PEEC members look to him for his advice.

In 2002, he joined the PEEC Administrative Committee as Secretary and continued to serve as its Vice-Chair and Chair for a continuous span of 6 years. In these roles, he provided unique leadership roles in serving the PEEC members. Because of his invaluable contributions to education, the members seek his advice all the time. He has thus become a natural mentor for both senior and junior faculty. I feel fortunate to be associated with him in this endeavor.

He has been a strong champion of the *Midwest Power Symposium* since its inception in 1969. He then made a case of expanding this event to all interested participants in the North American continent. The name of the event was changed to *North American Power Symposium* (*NAPS*) which is held every year around October. The symposium alternates between the east and west sides of the Mississippi River in a university that promotes power engineering and makes a strong case for holding the event. About 150 to 200 students and faculty participate in this event. Only students are allowed to make paper presentations and faculty members provide needed advice. This event provides an opportunity to learn of the latest research and educational activities happening throughout North America.

Professors Charlie Gross, Pete Sauer, and I joined Heydt in a paper on promoting the power engineering activities in 1999 NAPS as leaders of PEEC [18]. The initiative for developing the paper came from Heydt again. We subsequently published the article in the *Power Engineering Review Magazine* [19]. The following year Heydt proposed that we survey and collect all *High Impact* papers in power engineering since electricity was invented. Subsequently all PEEC members were asked to participate in a balloting process to select the top five papers. Heydt took a leadership role in the balloting process. These were presented in 2000 NAPS at the University of Waterloo [20]. A unique feature of these papers presentation is each paper was presented as though the original author would have presented it. All the five presenters found proper attire to make the event truly unique. Thanks to Heydt for the energetic role he played in promoting power engineering education!

8.3.2 Commitment

Jerry Heydt has always been deeply committed to providing a very high quality education so that his students, both undergraduate and graduate, will be placed in challenging career positions. The proof is in the pudding! After more than four decades of his commitment, his students are very highly placed and they are well recognized both in academe and in industry. Pete Sauer is one excellent example of such a success!

He has dedicated his career to power engineering education by constantly improving the quality and content of the courses he offered both at Purdue and Arizona State Universities. He paid particular attention to constantly refining the content of the courses he offered. He took deep interest in updating the curriculum content throughout his career. His commitment is evident in the quality of the two text books he published: one on *Computer Methods to Power System Analysis* [21] and the second one on *Electrical Power Quality* [4] already mentioned above. It is not an easy task to develop text books that have now become classics.

8.3.3 Reforms

Based on my informal discussions with him, I offer the following ideas for reforming the power and energy education to meet the demands of the twenty-first century.

- Offer self-paced undergraduate education via on-line courses.
- Offer Graduate Curriculum Development via on-line courses.
- Develop a series of smart grid courses.
- First step: Offer one or two introductory courses on smart grid.
- Have NSF, ECEDHA and IEEE provide the impetus and initiative.
- Bring experts to plan and develop the graduate courses using the latest IT technology.

Some of the topics that deserve consideration in making both undergraduate and graduate curricula relevant and meaningful are:

- Cyber and Physical Security (CPS)
- Smart Grid Basics
- Integration of Distributed Energy Resources including Storage
- Power Quality
- Reliability
- Optimization and Control
- Communications and computing (big data analytics)
- Microgrids
- Markets
- System Resiliency
- Sensors and other new technologies

I conclude this section with the hope that power engineering education will play a dominant role in shaping the future of the world for the betterment of the society.

8.4 Outreach

8.4.1 *Power Globe*

Heydt is the founder of the Power Globe as the e-mail forum. This is evident from browsing the internet on the *History of Power Globe* [22]. To quote: "In 1989, Heydt, then at the National Science Foundation, put together a list of e-mail addresses taken from the NSF database known as the *HP System*. This was a list of reviewers for the entire foundation. The names used were those who had participated in an NSF review in the power engineering program. The *HP system* was fraught with problems: many e-mail addresses were outdated and the list was poorly maintained. The idea of an e-mail bulletin board was described at the 1990 PES Winter Meeting and several useful suggestions were received on how to make use of the list.

At the 1990 PES Summer meeting, this was discussed further in connection with the possibility of forming a University Research Subcommittee of the Power Engineering Education Committee (PEEC). The idea was to use the subcommittee—and potentially the list of names and e-mail addresses—to network the community in matters of power research. Talukdar of Carnegie Mellon University assisted in this matter and volunteered to use the computer at CMU as a host. The number of names on the list was in the 25-50 range, and all were at universities. The machine at CMU was called the Globe machine, and the name Power Globe was given to the list. Talukdar, assisted by Sauer of the University of Illinois and Heydt—then at Purdue—put together the Power Globe.

Heydt composed the rules and edited the list; Talukdar made sure the computer implementation worked. There were a lot of problems since list editing was manual and often fell behind schedule. There were many disgruntled users who wanted their names added or removed in a timely way.

In about 1995, Ramesh, then a graduate student at CMU, was recruited by Talukdar to make sure that the list of names was maintained smoothly. This worked well initially, but the list grew steadily in size, and it became clear that a manually maintained e-mail list was not practical. The main additions to the list in this period were industry and government people on a worldwide scale. Ramesh completed his Ph.D. and left CMU for the Illinois Institute of Technology, where he continued to remotely maintain the Power Globe. This worked for a time, but maintenance of the list was still time-consuming.

In PEEC Research Subcommittee discussions, it was suggested that the operation of Power Globe might be streamlined by utilizing a more advanced listserver. North Dakota State was contacted in regards to the use of their listserver, which hosts over 1,500 e-mail lists and employs software that is in common use at many universities. Stuehm of NDSU investigated and found that it would be possible for Power Globe to be hosted there.

On August 7, 1996, Heydt transferred the existing subscriber list to NDSU and the present listserver-based Power Globe was created. Mork has been the list owner, coordinating the configuration of the list and providing subscriber maintenance. The list owner task will be rotated among Power Globe Working Group members in the future.

Operating experience to date has been positive, with high reliability of service. Since most subscribers can now take care of their own subscription on a self-serve basis, maintenance has been minimal."

8.4.2 Technical Paper Reviews

Another example of Heydt's outstanding commitment to service is his role as a reviewer for papers and publications. I had the first hand opportunity to observe this during 2004–2007 when I was Vice-President for PES Publications. In order to get timely review of the Transactions papers, the Editors-in-Chief surveyed the performance of more than 500 reviewers and found that he was the best reviewer in terms of the timeliness and quality of reviews. While the Editors provided 3 weeks to complete the reviews, Heydt typically took 2–3 days to complete the reviews. I wish everybody had followed his model.

These two outreach cases clearly demonstrate his dedication to the service of professional communities.

8.5 Summary

The essence of this chapter is to highlight the legacy of Professor G. T. Heydt to Power Engineering research, education, and outreach through a few cases of our interaction in the past four decades. We collaborated chronologically in the following areas: six-phase power transmission, power quality, modeling of highly nonlinear loads, and advanced generation technologies. I endeavor to summarize the remarkable contributions Professor Heydt made to these areas. I was, and am, indeed fortunate to interact with him because I have learned a lot in this process. I consider him one of the top leaders in engineering education in general and in power systems in particular. I sincerely hope that we will have many opportunities to interact in the future. I wish him a long and successful professional career so that many students and colleagues can benefit from his contributions yet to come.

Acknowledgments I wish to thank Professors Elias Kyriakides, Sid Suryanarayanan, and Vijay Vittal for inviting me to the event to celebrate the contributions of Professor Heydt and for giving me this opportunity to develop this chapter. I am highly thankful to Padma Venkata, my wife, lifelong partner and English Professor for editing this chapter and many numerous papers in my entire professional life.

References

1. Bhatt NB, Venkata SS, Guyker WC, Booth WH (1977) Six-phase (multi-phase) power transmission systems: fault analysis. IEEE Trans Power Appar Syst 96(3):758–767
2. Fortescue CL (1918) Method of symmetrical co-ordinates applied to the solution of polyphase networks. Presented at the 34th annual convention of the AIEE (American Institute of Electrical Engineers) in Atlantic City, N.J. on 28 July 1918. AIEE Transactions, vol 37, part II, pp 1027–1140
3. Clarke E (1950) Circuit analysis of AC power systems, vol I. Wiley, New York
4. Heydt GT (1991) Electric power quality, 2nd edn. Stars in a Circle Publications, Indianapolis, IN
5. Heydt GT, Venkata SS (1991) A workshop on research directions in electric power quality. National Science Foundation $12,047
6. Venkata SS, Heydt GT (eds) (1991) Proceedings of the NSF workshop on electric power quality, Grand Canyon, AZ, 130 p
7. Domijan A, Heydt GT, Meliopoulos APS, Venkata SS, West S (1993) Directions of research on electric power quality. IEEE Trans Power Deliv 8(1):429–436
8. Heydt GT, Meliopoulos A, Venkata SS (1994) Short courses on power quality offered to Tacoma City Light, CIA, Puget Sound Energy
9. Venkata SS, Heydt GT (May–Dec 1997) An EMTP module for nonlinear loads that cause bus voltage flicker. Electric Power Research Institute, $130,000
10. Venkata SS, Heydt GT (1998) An EMTP module for electric loads that cause bus voltage flicker. Final Report to Electric Power Research Institute, Palo Alto, CA, 100 p
11. Heydt GT, Venkata SS (1998/2000) Chaotic modeling of highly nonlinear loads. National Science Foundation, $115,000
12. O'Neill-Carrillo E, Heydt GT, Kostelich EJ, Venkata SS, Sundaram A (1999) Nonlinear deterministic modeling of highly varying loads. IEEE Trans Power Deliv 14(2):537–542
13. Jang G, Wang WG, Heydt GT, Venkata SS (2001) Development of enhanced electric arc furnace models for transient analysis. Elec Power Compon Syst 29(11):1061–1074
14. Jang G, Wang W, Heydt GT, Venkata SS, Lee B (2001) Development of Enhanced Electric Arc Furnace Models for Transient Analysis. Electric Power Components and Systems, 29:1061–1074
15. Venkata SS, Heydt GT (eds) (1993) Special issue on advanced power generation technologies. Proceedings of the IEEE, vol 81, no 3, pp 315–485. (15 papers and Scanning the Issue)
16. Heydt GT (1993) An assessment of ocean thermal energy conversion as an advanced electric generation methodology. Proc IEEE 81(3):409–418
17. Ren J, Venkata SS, Sortomme E (2014) An accurate synchrophasors based fault location method for emerging distribution systems. IEEE Trans Power Deliv 29(1):297–298
18. Gross CA, Heydt GT, Venkata SS (1999) The IEEE power engineering society – power engineering education activities in promoting the power engineering education. Proceedings of the 1999 North American Power Symposium, San Louis Obispo, CA, pp 267–272
19. Heydt GT, Venkata SS, Gross CA, Sauer PW (2000) Promoting the power engineering education through the IEEE power engineering society. IEEE Power Eng Rev 20(1):15–21
20. Heydt GT, Venkata SS, Balijepalli N (2000) High impact papers in power engineering, 1900–1999, North American Power Symposium, University of Waterloo, Waterloo, Canada, pp P-1 to P-7
21. Heydt GT (1986/1912) Computer analysis methods for power systems. Macmillan Publishing Company, New York, NY
22. History of Power Globe. http://www.ece.mtu.edu/faculty/ljbohman/peec/globe/index.html

Chapter 9
The Power Engineering Workforce in Washington and the Pacific Northwest: Opportunities and Challenges

Alan Hardcastle, Kyra Kester, and Chen-Ching Liu

9.1 Introduction

The shortage of the power engineering workforce has received great attention nationally in recent years. To facilitate a national strategy to address this issue, the National Science Foundation sponsored the Workshop on the Future Power Engineering Workforce on November 29–30, 2007. The forum was cosponsored by the North American Electric Reliability Corp. (NERC), IEEE Power and Energy Society (PES), and the Power Systems Engineering Research Center (PSERC). A coauthor of this chapter, Dr. Chen-Ching Liu, chaired the organizing committee of the Workshop. Leaders from universities, industry and governments developed a number of recommendations:

- Create a single, collaborative voice on solutions to engineering workforce challenges.
- Strengthen the case for extraordinary efforts to build, enhance, and sustain university power engineering programs.
- Envision the future challenges in electric energy supply and demand and develop an image that will increase interest in power engineering careers.

A. Hardcastle, Ph.D. (✉)
Washington State University Energy Program, Washington State University, Olympia, WA, USA
e-mail: hardcast@wsu.edu; alanh@wsac.wa.gov

K. Kester, Ph.D.
Social and Economic Sciences Research Center, Washington State University, Olympia, WA, USA

C.-C. Liu, Ph.D. (✉)
Energy Systems Innovation Center, Washington State University, Pullman, WA, USA

E. Kyriakides et al. (eds.), *Electric Power Engineering Research and Education*, Power Electronics and Power Systems, DOI 10.1007/978-3-319-17190-6_9

- Stimulate interest in power engineering careers and prepare students for a post-high school engineering education.
- Make the higher education experience relevant, stimulating, and effective in creating high quality and professional power engineers.
- Encourage and support increased university research to find innovative solutions and to enhance student education.

Following this workshop, with leadership from IEEE Power and Energy Society, a US Power and Energy Engineering Workforce Collaborative was established.

The Collaborative Management Steering Committee indicated that over the next 5 years, about 45 % of engineers in electric utilities would be eligible for retirement or could leave engineering for other reasons. The Committee also concluded that these losses would create a need for over 7,000 power engineers by electric utilities. In addition, two or three times more power engineers may be needed to satisfy the needs of the US economy. The first item of the priority goals identified by the Collaborative was to "double the production of undergraduate and graduate students in power engineering." Further background information can be found in references 4 and 5 of this chapter. The remainder of this chapter reports the results of a survey of the power engineering workforce conducted by Washington State University.

In 2008, the Washington State University (WSU) Energy Program completed a regional labor market and workforce study of electric power employers. The study collected data regarding new hiring, anticipated retirements and replacements, hiring challenges, and workforce education needs [1]. The study findings mirrored national predictions about the aging utility workforce, looming retirements, population trends and other factors that were predicted to create considerable labor and skill gaps in the electric power industry.

That initial study focused primarily on technical craft occupations: operators, mechanics, electricians, technicians and line workers. The engineering workforce was not a focus of the original study.

Since that time, the electric power industry has continued to transform how it generates, transmits and distributes electric power through the application of advanced technologies and processes, and other smart grid innovations. At the same time, the deepest recession since the Great Depression contributed to weaker short-term demand for electric power and delayed departures by retirement-eligible employees due to economic uncertainty and weakened retirement portfolios. How those two factors affect the power engineering industry and workforce were not well understood.

9.1.1 Purpose

This project was launched to provide current, systematic data on engineering employment to identify the labor market and workforce challenges in the Northwest, and particularly in Washington State. This study sought to find answers to the following questions:

- Have industry restructuring, new technology and the recession reduced the need for new hires or expanded demand in specific occupations and sectors?
- What are employers' estimates of the need to replace experienced power engineers due to retirements?
- What gaps do employers' anticipate, and what new succession plans or strategies do they have for filling these gaps?

Employer responses to these questions were collected to provide useful information for power engineering programs, faculty and students.

9.1.2 Methodology

To address these issues, the WSU Energy Program supplemented the 2008 research specifically for electrical power engineers, leveraging a portion of the research data collected for an expanded regional update of the 2008 study [2]. The project reviewed existing research and collected new data directly from a sample of Northwest employers. The combined quantitative and interview data were used to generate near- and longer-term forecasts for new employment and replacement of retirees, and strategies for filling key skill gaps.

To balance information from employers at multiple levels, the study gathered information by survey and by direct interview. As shown in Table 9.1, 18 Northwest energy companies from Washington, Oregon, Idaho, Montana, and Utah—representing a mix of utilities and energy service companies of different types, sizes, and geographic locations—were included. All participants were assured confidentiality.

The selection of the organizations was not based on statistical sampling procedures and the results cannot be reliably generalized to the electric power industry as a whole. The organizations represented the concentration of the industry in Washington, but many provide employment in nearby states, reflecting the regional nature of the labor market. The organizations include utility and consulting engineers, and those engaged in power system design, generation, transmission, and distribution.

Survey topic areas included:

- Total employment and current employment for power engineers.
- Current job vacancies, employment forecasts for new hires, and retirement replacements for power engineers.
- Succession planning and related strategies.
- Current and future training needs for power engineers.

The next section of the report includes:

- A brief summary of the major changes in the industry since 2008.
- Associated workforce and education-related trends.
- The reactions of representatives of the Northwest electric power industry.

Table 9.1 Participating employers and total employment

Employer	Total regional employment
Avista	1,672
Bonneville Power Administration	3,089
Chelan County PUD	643
Grant County PUD	721
Grays Harbor County PUD	152
Idaho Power	2,081
Incremental Systems	7
Northwestern Energy	1,428
PacifiCorp	6,251
Pacific Northwest National Labs	4,500
Portland General Electric	2,547
Puget Sound Energy	2,981
Schweitzer Engineering Labs	2,030
Seattle City Light	1,801
Snohomish County PUD	1,044
Tacoma Power	843
Transalta	296
US Bureau of Reclamation	1,093
Total employment	33,179

9.2 General Changes: 2008–2013

Two key factors have driven change in the Northwest electric power industry since 2008. The first consists of the continued modernization of the electric power infrastructure, particularly the implementation of technological innovations such as those included in smart grid installation. These include the technical and procedural changes driven by regulation and those created by continuing advances in technology. Technological changes are altering the industry in terms of power production and delivery, and changing the tools and procedures employed in all modern business. Customer expectations, customer skills and access to home technologies are changing the way that utilities do business just as deeply as they are changing businesses like banking and retail trade.

The second set of changes results from the economic recession that began in 2008. Despite official recovery, the recession's effects continue to reverberate in varying degrees around the nation. In parts of the Northwest, it may seem that the recession is well behind us, but the recovery has been deeply erratic, varying widely among communities and industries, as evidenced by power consumption and employment patterns.

For power engineers, then, changes come from two directions: from the change in the science of the power industry and from change in its business environment.

9.2.1 Regulation and Technology Changes

Much of the regulatory change still affecting Northwest utilities arises from the massive blackout in the Northeast in 2003. The binational, 3-months investigation that followed concluded that the blackout had been caused by a combination of human error and equipment failures. The final report recommended far-reaching changes to reduce the chance of repeating such a widespread event. These included replacing the voluntary standards for industry reliability that were established by the North American Electric Reliability Council with standards that were mandatory and enforceable. When Congress passed the Energy Policy Act of 2005, it expanded the authority of the Federal Energy Regulatory Commission (FERC) and required it to request, approve, and enforce new reliability standards for the new North American Electric Reliability Corporation (NERC) [3].

Since that time, the federal government has invested $4.5 billion in federal stimulus money toward the construction of a smart grid, enabling utilities to add hundreds of advanced grid sensors and millions of smart electric meters, which help power companies keep near real-time tabs on the state of the grid.

In fact, the increased demand of escalating regulation was a consistent refrain from those interviewed for this report. These regulatory responsibilities had a big impact on utility staffing and personnel. One common result of complying with these requirements is to combine the traditional core activities of an electrical engineer with those of computer engineers (see further discussion below).

Additionally, some of the technological changes arising from modernizing the grid required current workers to be retrained to ensure accurate installation, maintenance, and support. Adding new forms of electricity generation while utilities also investigate alternative energy sources can increase the need for additional training, as do the enhanced information and safety components of smart grid technologies.

9.2.2 Workforce Issues

Modernizing an infrastructure as widespread as the nation's electrical power system—which is critical to the nation's economy, national defense, and the general population's sense of well-being—is fraught with political and economic difficulty. Workforce issues are only one aspect of this complex undertaking, but these issues are critical because human error played a significant role in the power failure of 2003.

9.2.2.1 The Aging Workforce

Compounding the need for employees who are adept at managing new technologies is the imminent threat of widespread retirements. In fact, retirements have been a critical issue ever since the blackout, impelled by a National Science Foundation workshop in

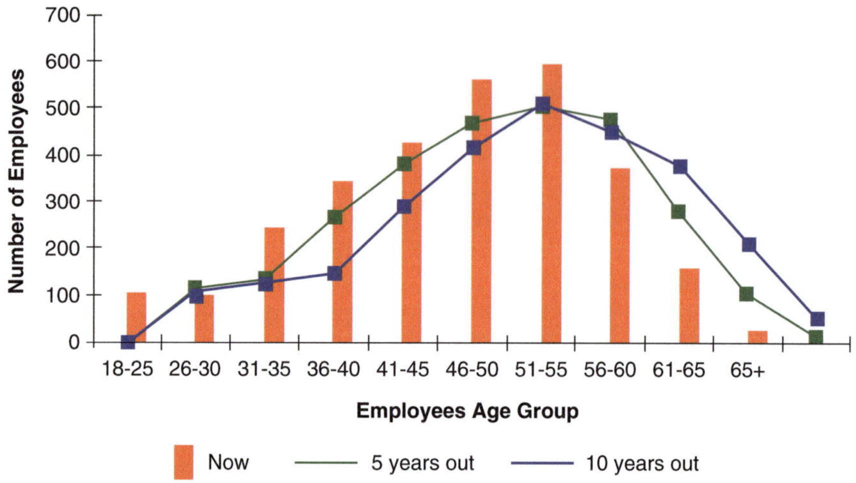

Fig. 9.1 The aging utility workforce. Source: Ray, Dennis and Bill Snyder. "Strategies to Address the Problem of Existing Expertise in the Electric Power Industry." *Proceedings of the 39*th *Annual Hawaii International Conference on System Sciences*. January 2006

2007, when the Power & Energy Society of the Institute of Electrical and Electronics Engineers (IEEE) founded the US Power and Energy Engineering Workforce Collaborative (PWC) [4]. Its charge was to strengthen the US power and energy workforce. In a 2009 report, the PWC noted that approximately 45 % of US electric power engineers would be eligible for retirement or could leave engineering for other reasons in the subsequent 5 years (see Fig. 9.1) [5]. A corroborating survey conducted by an industry consortium, the Center for Energy Workforce Development (CEWD), in 2008 found that this decline could reach 40–50 % by 2013 [6]. A more recent survey by CEWD found that while the percentage of potential engineering replacements had declined to around 38 %, due in part to the recession, this nonetheless represents the potential replacement of 10,600 engineers between 2010 and 2015 [7].

These general concerns about the age of the utility workforce are not lost to employers in Washington State and across the Pacific Northwest [8]. Figure 9.2 shows that among the age cohorts for all industries in the Pacific Northwest states combined, 37 % of the workforce is made up of employees under age 35 and just 20 % of the workforce in all industries is age 55 or older. In contrast, for each state's utility sector, just 20 % or fewer of employees are under 35, while 30 % or more are 55 and older. More broadly, over 60 % of regional utility workers are now 45 years of age or older.

9.2.2.2 University Programs

Unhappily, while the need to prepare new workers was increasing, many university power engineering programs were weakening. This was largely due to the

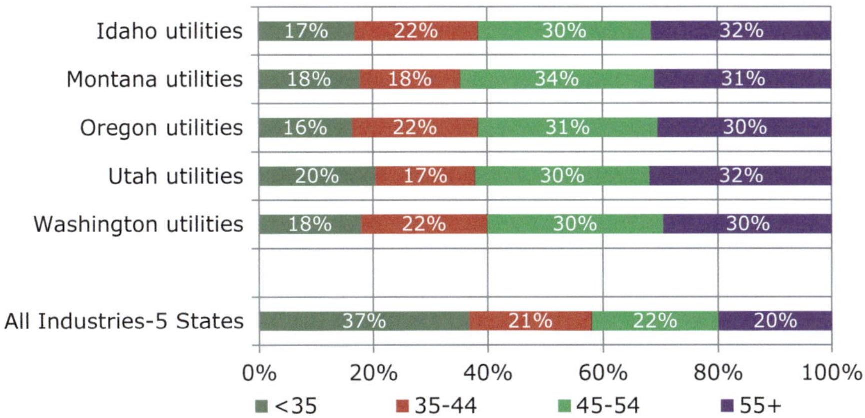

Fig. 9.2 Age cohorts and percentages by state: utilities versus all industries (five states combined)

increasing popularity of other electrical engineering specialties, a decline in research funding to support graduate students, and low student interest in science, technology, engineering, and mathematics in general.

Additionally, the 2009 PWC report indicated that approximately 40 % of power engineering faculty at US universities would become eligible for retirement within the next 5 years, and 27 % were expected to retire [9]. The gap in the number of engineers available for hire cannot be filled without faculty in place to train them.

This concern is echoed by many industry analysts concerned that the pipeline to replace experienced engineers is not as dynamic as needed, largely because so many professors in university power programs face retirements and many programs have severely limited faculty replacement allotments.

The PWC report noted that overall enrollment of university students in power and energy engineering courses was increasing. The report speculated that the increase resulted from a growing interest among young people in renewable energy systems and green technologies. Yet the report also pointed to survey results showing that the overall number of students interested in electrical engineering was declining, and that the number of students completing power engineering degrees would need to double over the next 5–8 years to meet future employment demand from utilities and other industry sectors. A shrinking pool of electrical engineering students limits the future supply of new power engineers.

All of the industry representatives who were interviewed for this study indicated some level of interaction with the universities where they most frequently recruit new employees. The industry representatives generally understood the need to have good working relationships with engineering programs and their students. How closely they were engaged with campuses varied widely, from those who went only for recruiting and informational opportunities, to those with ongoing curriculum discussions with faculty. Most had what they considered a good relationship with between three and six universities where they routinely hired graduates.

Describing the specific challenges they faced when recruiting power engineers, many industry representatives replied that engineers with experience in power planning and operations were the most difficult to find. Industry needs engineers who understand the nature of compliance with new and ever-changing regulations, including planning for upcoming changes. Too many applicants had too little experience, particularly in power generation, and no experience in a large power plant. Others found that the applicants had too little understanding of the challenges of providing urban power service.

Nearly all industry representatives found few applicants with experience that extended beyond the classroom. As one group of engineering managers noted, "We can find HVAC-experienced engineers, but not engineers with large energy plant, power generation experience."

Some utilities have benefited over the past few years from a stable workforce, as retirement-eligible employees chose to defer their departures due to the recession. Now, as recovery gains traction, employers are finding that they have acute hiring needs but a shortage of qualified applicants and few recruiting experiences to guide them.

Employers also need supervisors for mechanical and electrical engineers across the electrical grid, but cannot find applicants with relevant experience. Some utilities that are now using recruiters are finding that advertisements are attracting entry-level applicants, but not experienced applicants.

Employers also report compensation to be a substantial issue through the entire salary range, although there are industry variations. Federal employers found themselves at a disadvantage for the first few years of an engineer's employment due to very low starting salaries, but their attractiveness increased after a few years with the steady "step" increases of public personnel systems. Other employers seemed able to set starting salaries at a competitive level, but were not able to advance salaries at the same rate as industry partners, contributing to the churn of mid-level engineers.

9.2.2.3 Effects of Recession

For a while, the economic recession stalled many retirements and made employees less likely to change jobs. Both factors reduced the number of potential replacements required in the short term. The CEWD survey in 2011 illustrated the change [7]:

- More engineers overall.

  Even during the recession, utilities continued to hire engineers—a 3.6 % increase nationally—even while the size of the industry workforce decreased overall by 11,000 jobs in the 2 years from 2009 to 2011.
- Fewer employees leaving for reasons other than retirement.

  "Forecasts for annual attrition for reasons other than retirement decreased over previous years from an average of 5–2.2 %, most certainly a result of the (recession)."

- Fewer retirements, which move the looming retirements forward, but increasing the potential abruptness of their loss.

  "The workforce continues to mature—the average age of the Electric and Natural Gas Utility workforce has increased from age 45.7 in 2006 to 46.1 in 2010. Comparisons by age groupings show that the number of employees between the ages of 18–27 has decreased while the number of employees age 53 and above has increased, reflecting both the number of mid-career hires and the number of employees who are waiting to retire."

Other Issues

In addition to the effects of recession, many parts of the energy industry remain affected by the federal budget cuts and sequestration. This is particularly true of federal organizations in the Northwest, such as the Bureau of Reclamation, part of the Department of the Interior that runs Coulee Dam. These cuts appear particularly damaging to internships, which are commonly used as a way for the power industry to attract, initiate and select potential employees from among engineering students. Federal internship programs, such as the one at Coulee Dam, have been heavily restricted by sequestration.

These conditions form a backdrop to the information gleaned in this study from representatives of the Northwest power industry. The next section presents analyses of employment-related data, future forecasts, and commentary on the workforce conditions these employers face.

9.3 Results

This study inquired specifically about employment and workforce issues that affect electric power employers and engineers across the Pacific Northwest. These include:

- Current demand and vacancies.
- The aging workforce, retirement forecasts and industry responses to it.
- Recruiting new workers and future demand.
- Supplementing the skills of new workers.
- Supplementing the skills of experienced workers.
- Attracting youth to the industry.

It is important to note that this study is based on a limited employer sample, and the data do not account for vacancies, retirements or hiring estimates at other organizations. Thus, the data reported by employers likely understates the actual number of current and future employment opportunities available to power engineers across the region.

Table 9.2 Current vacancies and hiring expectations by occupational group, 2013

Occupational group	Number of vacancies	Number of vacancies employers expect to fill	Number of vacancies expected to remain unfilled
Operator	28	28	0
Mechanic	15	15	0
Electrician	39	25	14
Technician	24	22	2
Line worker	60	47	13
Power system operator	18	18	0
Power engineer	85	84	1
Customer service representative	51	50	1
Energy efficiency program manager	3	3	0
Total in these occupations	305	275	30

9.3.1 *Overall Demand*

The primary method for measuring overall demand in the industry is what employers report as their current "demand," based on their current openings and anticipated needs. To learn this, we surveyed employers in the electrical power industry in the Pacific Northwest. These employers reported their demand in a critical core of occupations; their responses highlight the need for power engineers.

As shown in Table 9.2, employers reported 85 current vacancies for power engineers, accounting for 28 % of total vacancies among the nine occupations studied. The 11 employers who had engineering vacancies also reported that, despite the many challenges of recruiting and hiring qualified power engineers, they expect to eventually fill all but one of the openings they have available. Although it was not a focus of this study, it is worth noting that the electric power industry provides very high-wage employment and compensation for power engineers is among the highest across all occupations [10].

9.3.2 *The Aging Workforce*

All the industry representatives interviewed for this study were asked about the retirements they anticipate and how they were planning for those retirements. In most cases, employers were well aware of the issue; most knew how many of their current engineers were or would soon be eligible to retire and they knew the imminent plans of those engineers. Table 9.3 illustrates employers' 5-year retirement projections across several occupations, including power engineers.

Table 9.3 Anticipated retirements in FTEs per occupational group, 2013–2018

Occupational group	Number of retirees projected	Percent of current workforce expected to retire (%)
Operator	152	14.6
Mechanic	150	16.4
Electrician	251	26.0
Technician	158.5	18.6
Line worker	386	18.2
Power system operator	66	17.0
Power engineer	210	20.5
Customer service representative	144	8.8
Energy efficiency program manager	38	22.9
Total	1,555.5	17.0

Over 20 % of current power engineering workforce is expected to retire over the next 5 years. Further, the 15 employers who estimated future retirements and indicated replacement values for power engineers indicated that they expect to replace all openings.

Naturally, employers realize the potential damage caused by losing so many senior staff in a short period. In fact, *no employer reported being unaware of the looming problem.* Many have developed strategic plans for replacing their veteran staff as they retire. How much planning—and how well it has been absorbed at the operational level—varies, as evident in the comments of industry representatives with whom we spoke.

At one company, for example, a unit manager stressed, "We have a succession planning culture. We have had for about 10 or, really, 15 years. We always try not to get caught with retirements or people leaving and us not knowing how to backfill those positions."

Another employer, for whom senior staff members are particularly crucial, reported, "We have a hiring plan approximately 5 years in advance of need. It's updated annually, and it's fairly seamless. And for senior staff we have a succession plan. We target those, but of course everyone will be replaced at some point."

More often, however, while employers were thinking about the issue and most were aware of their employees' ages, their approaches to the problem varied in different units or were largely ad hoc. Describing a common condition, one representative said, "It's really up to department managers. We do have a system for filling critical needs, but it's mostly up to department managers to determine how to do this. It's up to them."

Employers frequently referred to the importance of their internship programs. A representative response was, "A few years ago, we started a program of hiring seven interns each summer to evaluate and expose them to the industry. The number had fluctuated before with the budget, but for 3–4 years it was seven consistently.

We followed that with a plan to hire young engineers, even creating positions for them to backfill potential retirees. We prefer to hire to a specific position."

Many employers rely on grooming interns as a way to fill openings, although filling entry-level positions does not address the retirement of a senior engineer unless, as many noted, the entire spectrum of staff is also prepared so they can be promoted to fill senior positions.

A few managers, primarily with larger employers, also mentioned that their succession policies have evolved since the issue of retirements arose. One manager noted, "Our succession planning has changed and evolved, of course, as staffs have changed and different people have been involved. And it is driven by the needs of different sectors (of the utility), as some parts have more needs than others." These variations may be practical, but they make it more difficult to observe and report a singular approach to addressing retirements.

At several utilities, section managers admitted that while their human resources staff might have plans on file, operational reality could be different and usually varied with each retirement. When faced with openings and no applicants, what did they do? Some changed the work so it could be done by less experienced engineers. Some hired new graduates and trained them. While these employers have investigated hiring from other industries, only a few have done so successfully. Most reported that the work was too dissimilar. "It works for other trades, such as electricians, but not for engineers."

9.3.2.1 A Recycled Asset: Retire/Rehire

One obvious way to address retirement is to hold on to the employee who is eligible to retire for as long as possible, providing incentives for deferring departure. Another strategy is to utilize the employee after retirement to help address the longer-term challenges associated with the transition of skilled employees. Among the targets are retirees who can be rehired on a part-time and/or project basis or consulting organizations that have hired retirees.

Because retirees offer the benefit of institutional knowledge, many employers report using them to offset the lack of experienced engineers for mid-career positions. None of the respondents reported using retirees as a substantial part of their workforce; rather, many seemed only to utilize particularly skilled or unusually experienced retirees. A common example: "We use retirees just a little bit. Two examples are a former employee who is a mechanical engineer and one who was always a consultant to fill vacancies on an as-needed basis."

A few employers described the relationship between consultants and retirees. When senior engineers retired, skilled mid-level people were not always available to fill those positions. As one employer explained, "You hope to have somebody you've been working with who can step into the role. But if not, then we recruit—advertise in the larger engineering community first. But if that expertise is just not out there, then we look for temporary solutions until we can train up an internal employee. Sometimes we can use a retiree on a temporary basis."

At least one employer appreciated the contributions of retired employees so substantially that they were given emeritus status and a workspace was maintained for them. As a result, said this employer, "...Many stay around half days. They are not rehired, but also not doing regular work. They add credibility to what we do."

9.3.2.2 Hiring Contractors

Many employers reported utilizing consultants to help address staffing needs. "We use contractors quite a bit. We have contractors with whom we work regularly and they send us engineers from all over: Florida, Chicago, Denver. We use them for engineering, as project managers, for support positions."

Frequently, contractors were relied on to meet temporary needs for which permanent staffing was not required. A common example was the need to make changes in workplace processes to comply with regulations. Once these changes were made by consultants, the ongoing duties were absorbed by regular staff. "We have a lot of work right now with license implementation. There's just no way we can do all the design work for the floating surface collectors, for the hatcheries, for that type of work, so we are trying to stay involved and managing (the contractors) with our own in-house engineers and doing as much design as we can, but there is just a level we cannot accomplish. So, yes, we are hiring consulting engineers. And they are having similar (staffing) problems!"

In other cases, costs motivated the hiring of consultants. Many employers reported that consultants were hired more cheaply than permanent staff. One employer spoke specifically of hiring a consultant in order to use one of their staff, who was later hired directly when the budget permitted. Others mentioned similar examples of hiring staff from consultant companies. "I hire the best I can find, I don't compromise. If you do, you pay for it later. The consulting firm appreciated having the work we gave them, but they didn't appreciate me taking most of their best people!"

One reason cited for the "poaching" of experienced staff was the cost of mentorship and training, which had suffered budget cuts in recent years.

> We used to hire a bunch of new engineers, rotate them around the company, but then hiring leveled off. We may get back to that model because we have to—we can't seem to find (experienced engineers) on the market—but for now, the financial realities just don't allow for us doubling-up on hiring (meaning, hire and put with mentor for 3–6 months, for example, to train and do knowledge transfer). We just can't double up on hiring to do student-mentoring anymore.

9.3.3 Attracting and Preparing New Power Engineers

Extending relationships with retirees is not a permanent solution to a personnel shortfall, and the situation is likely to worsen soon. As CEWD noted in 2011, the current prolonged employment of senior workers is likely temporary because the

Table 9.4 Projected staffing change in FTEs per occupational group, 2013–2016

Occupational group	Net growth in FTEs	Total current employment in occupation	Percentage change next 3 years (%)
Operator	2	1,039	0.2
Mechanic	18	912	2.0
Electrician	6	964	0.6
Technician	1	854	0.1
Line worker	1	2,120	<0.1
Power systems operator	0	388	0.0
Power engineer	21	1,027	2.1
Customer service representative	0	1,635	0.0
Energy efficiency program manager	1	166	0.6
Total new employment	50 FTEs	9,105	0.5

economic recovery may make retirement feasible once again. That may, in turn, cause employers to begin attracting mid-career engineers away from each other. No matter how much churn occurs, new engineers are needed and the supply of power engineers in the pipeline becomes more critical.

As part of this study, employers were asked about their anticipated hiring in the next few years in nine occupational groups. The results appear in Table 9.4. Although projected new hiring among these employers is very modest, the anticipated hiring is highest for power engineers.[1] When combined with current vacancies and especially employers' estimates for retirement replacements, the data further emphasizes how critical the availability of new power engineers will be.

9.3.3.1 Recruiting New Engineers

In the study interviews, employers were asked specifically about how they recruited power engineers and the difficulties they faced doing so. Table 9.5 summarizes their responses.

Overall, employers reported that power engineers were the most difficult employees to recruit and hire. Several noted that the competition for experienced power engineers (mid-levels and above) was especially intense, and that younger engineers—especially electrical engineers with power experience—were frequently lured to other companies by higher compensation, better work conditions, or other factors. Employers noted that these young professionals tended to be very

[1] This finding is reasonably consistent with the existing state labor market forecast for Washington, which shows electrical engineers as an in-demand occupation with an annual growth rate of 1.9 % between 2010 and 2020. However, these data do not report specifically for power engineers. See: https://fortress.wa.gov/esd/employmentdata/reports-publications/occupational-reports/occupations-in-demand.

Table 9.5 Recruiting and hiring challenges

Position	Challenges
Power engineers	Requires 4-year engineering degree, usually electrical engineering
	Hardest to find
	Prefer power generation experience
	For smart grid work, need IT and computer science/software and automation knowledge and related skills
	Requires multi-state recruitments to secure qualified engineers
	Experience high turnover; many newer graduates, in particular, stay for only 3–5 years
	Lack of preparation requires lengthy in-house training progression

mobile, and that high demand for this occupation meant that qualified candidates had many employment options both within and outside of the energy industry.

9.3.3.2 Cultivating Relationships with Universities

Given these reported challenges, employers were also asked how they found their new, entry-level hires. Most immediately cited a small number of engineering programs they worked with regularly and recruited from most often. The University of Washington and Washington State University were frequently mentioned as the regional public universities producing the most engineers. Others named Seattle University, Gonzaga University, Idaho State, Portland State, and a small number of schools outside the region (Howard University, for example). And one employer, with a particular need for employees who would need to relocate to a very rural environment, preferred to recruit from Eastern Washington University and focus on students from rural backgrounds.

Two employers expressed related concerns about the intensely technical focus of engineering programs.

- One was concerned that there is no effort to address a key characteristic of Generation Y,[2] which is the desire to find their work meaningful and to contribute to change. The employer argued that this case could be made more strongly by faculty and through engineering programs. However, these programs are usually so heavily focused on the technical aspects of engineering that the personal values and societal benefits associated with engineering education are not adequately emphasized for the millennial generation. This deficit in engineering education is likely leading potential engineers to turn elsewhere for that satisfaction.

[2] In the U.S.A., Generation Y refers to individuals born in the 1980s and 1990s, comprising primarily the children of the baby boomers and typically perceived as increasingly familiar with digital and electronic technology. Generation Y is synonymous with the Millennial Generation.

- The other employer was concerned that over-emphasizing technical skills may push out candidates who could do the work required and might, in fact, bring more well-rounded characteristics to their jobs. These are characteristics the industry is actually looking for, such as having a team-based approach, interdisciplinary skills, project management skills, and broad knowledge of computer, IT, and communication software and applications.

9.3.3.3 The Role of Internships in Recruitment and Pre-training

Nearly all the industry organizations engaged actively in internships to attract engineers to the field, provide practical experience in the industry, and provide a first look at potential future hires. Those employers that did not hire interns were restricted by budget constraints.

Despite the important role that internships play, especially in light of the potential shortfall of workers, the number of engineering students for whom internships are available seems small. Typical responses included:

- "We will have two or three per year."
- "Typically we hire four interns. We screen an application, check GPA, etc. Then we interview, introduce eight applicants to staff, and we will hire four."

A few employers were able to do more:

> We hire 20 summer interns. They are work assignments and the interns apply like they would for a job, but we expect less of their qualifications. We provide a mentor, but nothing more formal than that, except that they should participate in our summer lecture series, as all staff do, which are 1-h lectures on a variety of topics. Any of us can sponsor an intern, so the focus is on what the project needs. And we also have graduate program internships of 3–6 months.

As might be expected, budget allocations determined the final number of interns. At least one employer found that pre-planning helped them identify how many interns they were likely to support. If they were planning to assign work from an ongoing project, then the funds already assigned to that project would cover the internship.

Those who mentioned recruiting from specific schools reported that, in their opinion, the strength of interns from the programs they targeted reflected the quality of the engineering program there. Many reported recruiting nationally, particularly those employers who expressed concern about recruiting minorities. Still, several also expressed dismay that successful recruitment of interns or new graduates did not necessarily correlate with successful retention, particularly for those employers with lower pay scales and challenging community environments.

One employer also reported working specifically with the Multiple Engineering Cooperative Program (MECP), typically sponsoring three MECP engineers per year.[3] "And we do that in generation, as well as substation design areas."

[3] Multiple Engineering Cooperative of Oregon universities, an exclusive program, sponsors 3, -6-months internships in the industry. Students require 5 years to graduate due to the internship, which is paid ($17/hour) Interns are given simple to advanced projects to do, and are given guidance. Interns include electrical, civil and mechanical engineering students.

Interns were usually placed throughout the company. Most, in fact, were "brought into specific departments" and matched to specific projects. "We hire interns every year—across the company, in many disciplines." The assignments varied, but all interns were required to get at least some field experience. In a few cases, interns were intentionally rotated through a variety of departments to extend their exposure, but most were "exposed" to different work areas rather than formally rotated. "They must also be exposed to other units, at least one, and take tours and participate in detailed project discussions."

In a few cases, the internship was a structured program, utilized across the company. In more cases, however, interns were hired as needed and used where projects could utilize them. "Historically these have been project driven, so we look for projects during the year that would suit an intern well. (We) might be looking at efficiency unit data, it might be inventorying asset management related, it varies, and if we have a solid project we bring an intern in. Usually one per year, two maybe last summer." For a few employers, that meant planning throughout the year to identify which projects could utilize an intern. But for most it meant a commitment to using interns who, once hired, were matched with available work.

According to the employers who were interviewed for this study, internships were not usually structured and depended on matching interns with appropriate projects after they had been selected. As one employer noted, "To my knowledge, we don't have a written plan that would describe what we hope the intern goes away with. We should, but we don't." Some utilities had specifically defined programs, focusing on corporate business practices, for example, but most engineering interns were simply assigned to do current project work. And most firms found the internships fruitful: "We have only ever once had an intern who clearly wouldn't fit in our culture. All our other interns, we've had no difficulty and have hired several."

Lack of formal program definition did not necessarily mean lack of forethought, however. Several employers described the intern hiring process as the midpoint in a planning/hiring/supervising continuum. "Interns are planned for ahead of time so the unit that will be getting one is part of the interviewing, takes responsibility for the intern, and takes ownership of delivering at least as much value to the intern as the intern might deliver to us."

While most organizations may be committed to recruiting and employing interns, the experience of each intern could vary widely, even when with a single employer.

Many engineering units plan for their summer interns throughout the year, examining projects for opportunities that might suit a student, and then requesting an intern for the work. Other worksites recruit interns, and then as they interview the applicants, determine where they might be most useful and most successful.

> We use interns in specific areas: substation engineering, protection and control, and planning/operations. We are targeting more juniors, and depending on the coursework they have taken—and if they have some power-related courses—we assign them. But we tell them where they are to be placed, who would be the mentor, and ask if that is what they want to do. We get their input into the assignments. The internship typically lasts 13 weeks.

> We do two batches: normally WSU students start early, in the middle of May; later, UW students come on as school ends.

So important are internships to students that some seek to repeat the experience. When employers were asked if interns were allowed to repeat, responses varied. One employer told us, "We've had a couple return. We don't have a formal policy to try to hire different people and spread the wealth (of opportunity) around, but there is some consideration of not hiring interns too early or too late in their studies, so there is probably one good year or maybe two when this is most valuable." And sometimes interns were not only extended, but were hired directly: "In two cases, we extended longer than summer and finally just hired them. I have a case like that now."

And the interns are right about the importance of the experience. Almost all employers who used interns reported a high likelihood of hiring those who proved themselves. As one noted, "Our internship is very successful. We have hired about 25 % of them, usually two out of every summer's seven."

Every industry representative with whom we spoke reported at least some difficulty hiring engineers. The most common difficulty was finding experienced engineers, and particularly those with 15 or more years of experience. Many employers found salaries a hindrance, particularly among public entities that did not feel competitive with private industry. This was particularly true when competing for PhDs or even engineers with master's degrees. Entry-level hiring seemed comparatively easier.

> Yes, we mostly hire younger and train. We prefer to find PhDs and industry experience, but it is rare. And only for 3 or 4 years are our salaries really competitive. So, we hire out of college, with all the experience on paper, no field experience. It doesn't cause any real problems, except that we have to provide that knowledge and make careful benchmarks for developing experience.

9.3.4 Supplementing the Skills of New Workers

Many employers have created specific processes for adding to the skill base of newly hired engineers. Most were not formal training programs; rather, they were typically unstructured training experiences, relying on current workers to provide varying degrees of oversight and mentorship. A few employers remarked that they found this requirement acceptable. They seemed particularly understanding of the breadth of material that a baccalaureate-level program must cover and sympathized with the need to make the bachelor's-level instruction generic enough to serve a diverse industry. They were not dissatisfied with the need to seek master's-level candidates for power engineering specializations, even accepting that graduates with master's degrees still required additional training in the specifics of their organization. As one employer noted, "Even with a PhD or master's, we find more expertise, but not experience."

Most new hires are young engineers. Although internship experiences were described as an asset for entry-level applicants, they were not prerequisites for hiring. Given the small number of internships apparently available, that would be untenable. Thus, most new hires arrive without much—if any—background in the industry. In fact, many industry representatives lamented the lack of industry exposure among new hires.

Therefore, in our interviews we inquired specifically about the preparation of new engineers. We asked about the technical preparation of graduates, but also about their general preparation for working in the electric power industry.

Employers indicated strong agreement that the graduates hired from Pacific Northwest universities and a select number of out-of-region schools were well prepared in basic engineering.[4] Some employers commented that the new engineers recruited from schools where they had cultivated relationships were well prepared. There was equally strong agreement, however, that new graduates were far less well prepared for work *in the power industry*. "Power system engineering under electrical engineering is not taught at enough schools," said one employer. It was generally observed, too, that this largely resulted from student choice. "EEs want (courses in) microelectronics, not power."

A few employers noted specific issues about internships that others did not mention.

- One observer noted that the issue of hiring, both for engineers and interns, is particularly American. "We have a great shortage of U.S. citizens available in the industry. It's not sexy. We look at the students of 11 different nations, countries paying them to come here to train and then return home. Two-thirds of my staff are foreign nationals, and it might be more except that security clearances require citizenship, at least in our setting, for which cybersecurity, federal and state security clearances are required. But recently we had posting with 16 mid-career applicants, and only one was a U.S. citizen."
- A related issue was candidate diversity. "We make specific recruiting efforts for women and minorities. It's just that the pool's not that deep, so when someone appears, they often leave here relatively quickly. The real problem is the general supply. They get multiple offers."

Most of the concerns about preparation for the electric power industry fell into two categories:

- Technical skills, including those particular to utilities and those likely required by many employers.

[4] One employer disagreed strongly, alleging that engineering programs generally were not preparing electrical engineers adequately. This employer hired nationally and reflected on the engineering graduates broadly. The reference was also made, generally, about older engineers rather than recent graduates. "We have seen EEs with degrees who don't know the basics of electrical engineering. This was (among) the older group, who studied certain aspects of EE but didn't know others." Their solution was to determine which universities had the strongest programs and recruit only there.

- Social skills, likely affecting any employment, but a strong shortcoming for the energy industry work.

9.3.4.1 Technical Skills Lacking Among Young Engineers

Employers described technical skills they saw as lacking among young engineers.

Power Industry Specificity

The most common complaint among the employers who were interviewed was the lack of exposure to and experience with power industry equipment, processes, and requirements. Many employers observed that "fundamental power system engineering principles" had shifted out of undergraduate programs and now resided in graduate-level studies. Most acknowledged that the electrical engineering undergraduate program was demanding in order to include all the core requirements, but the most senior interview participants felt that was acceptable. "When I was there, EE was way more credit hours than other programs, because you really needed the time (to cover those essentials). Things like basic power flow and sequence components training seem to be lacking at undergraduate levels. We're getting MSEEs now with that background, but I did most of that in undergrad courses."

Many noted that some graduates lacked a "broad, complete foundation in power system basics." One noted, "System protection and power flow are big, important topics to (cover)."

In related comments, employers observed that new graduates were "unfamiliar with the equipment. They haven't seen it before except in textbooks. They have textbook skills, but it takes them awhile to distinguish between bushings, insulators, and other equipment pieces." Other practical applications were also too rare. "Reading drawings; making drawings. They are taught theory, but not how to put together a contract preparation or how to put a design on paper for a bid."

Computer-Related Knowledge

Several employers considered the division between computer and electrical engineering too rigid. "Our electrical engineers need more computer-related knowledge and skills. That's where the industry is going. The electrical engineering core is not functioning unless the engineer has enough computer skill to know how an electrical control device is going to work. Otherwise, you are reliant on someone else."

They also found that while younger engineers were comfortable using computers and electronic devices, they tended "to trust computer tools too much. They assume the result is correct without doing a common-sense check."

Although these observations might seem contradictory at first, the uncritical reliance on computer-generated results may indicate a critical lack of expertise.

General Business Practices

For many, the most critical problem was the lack of basic understanding of the business world. On one level, the problem was lack of any kind of workplace experience, with the result that new hires had no idea how to behave in an office environment (see below). But even more critical was the failure to understand how engineering and the business of the organization intersected, and what responsibilities that entailed for them. Several strongly conveyed that their new hires knew nothing of the business and legal processes that are critical to the industry, and had no appreciation that those principles were "at least as important as their engineering products."

> "We have a really strong interest in not becoming extinct. (That means) producing power for attractive rates and with high reliability and, therefore, we have a need to talk at least as much in business principles as we do in engineering principles."
>
> Pacific Northwest employer

Others noted that, while some students had taken courses in engineering economics, somehow it did not prepare them for how decisions were made in a business environment. New engineers did not commonly deliberate questions such as "Is this cost effective?" or "Is it too radical for this customer?" according to the employers. As one employer reported, "Most of what we have to teach is in terms of contract research: "Who are our customers? What difference does that make?" Many of the employers we interviewed wondered why new power engineers lacked the ability to answer practical, useful questions such as these, which should be included among the skills they learn as part of their college degree.

Use of Data, Statistical Analysis, and Presentation Skills

In a somewhat related observation, employers found that new engineers were not prepared to use data to explain their recommendations, particularly in a manner understandable to the general public. They felt that "data analytics, that's getting to be a big and important function, big for our industry." And that "data management skills, pulling the data and presenting the data to non-technical people" was essential. "We have 825,000 customers, each one has a smart meter on their house and we're getting tons of data from them. We're getting tons and tons of data, and we need to make sense of it. Both data organization and management, and the analysis process" were deemed critical skills.

Respondents carefully distinguished this from computer skills, noting that what they needed from their power engineers was not what IT staff were accustomed to doing. "IT likes to think its function is data organization, data management, but they don't get the operational side of our business well enough to do it so that it is functional for us. We need to partner with them, but a lot of that will rest on our staff people: we need to extract (our own data), put it into usable format. IT should manage the data warehouse and big files, but we have to be able to manipulate and make sense of the data. (A generic IT person) doesn't know what to look for."

Other Communication Skills

Power engineers need to know how to "translate" technical data for consumption by the general public consumption, and for the trades and crafts employees with whom they work.

Communication skills were commonly cited as an issue, although they are for most employers of every kind of employee. Several referred to the introverted nature of many engineers, who would prefer to work alone or only with other engineers. Few new hires were accustomed to making presentations or even participating in group discussions. "Engineers (who) can talk and communicate seem to be few and far between."

The most important communication skills mentioned by these employers were related to communicating technical information to the public and to trades workers. Communicating with the general public was often about explaining new products and processes; communicating with other workers was related to providing instruction and persuading the other workers that the engineer had the right answer to a problem. For some employers, this was a vital issue: "Engineers have to convince workers of the solutions they've decided to implement."

The lack of communication skills hindered new engineers' career progression. Mid-level positions are often characterized by more shared responsibility and management duties. Without communication skills, engineers were less likely to rise in positions of leadership. "Here, we have mostly business leadership because it is hard to find engineers who are as socially aware as they are technically skilled."

Project Management

Other general workplace and career preparation skills that new engineers seem to lack involve project management. Even without personnel management responsibilities, power engineers were commonly assigned to run projects, yet many managers found their project management skills weak. "Budgeting, scheduling, and organizational skills generally" were often cited.

Basic Workplace Skills

A large number of employers pointed to even more basic skills that new engineers lack: very basic social skills and common workplace skills, often termed "soft skills." The employers talked about the nature and culture of the power energy workplace. "Utilities are conservative institutions, staid.... Some new hires are surprisingly rude, self-absorbed, and lacking in social skills. I think interdisciplinary studies would help them consider other points of view, other ways of being. They would go further professionally and have more career opportunities with (better) social skills."

Teamwork and Work Habits

Some employers reported personal knowledge that engineering programs were trying to increase students' exposure to teamwork through projects, but they wondered if the commitment was serious enough. They still found too many graduates who did not expect or want to work with others. "They are reluctant to work on team projects that involve other trades and engineering disciplines."

Many employers believed this lack of preparation for the realities of the workplace was a disservice to students and recommended including a basic workplace readiness component to engineering coursework. "(Some kind of) organizational behavior course (is needed. Preparation for) Engineers as part of a larger company is especially important for graduates without any significant working experience. Just a little primer would be an asset."

While few critics considered these work habits more important than basic engineering skills, many reflected on the demand placed on employers to develop these skills in new hires and the enormous advantage of job candidates who could demonstrate these workplace skills.

9.3.4.2 Teaching Industry-Specific Skills to New Engineers

Most employers had grappled with the best way to impart industry-specific skills in new engineers and arrived at the same conclusions:

- Provide mentoring by experienced employees.
- Assign new engineers to diverse projects.

Provide Mentorships

Faced with a common array of shortcomings in new engineers, most employers reported relying on more senior staff to help initiate new employees. Most referred to the system as "mentorships." Although the term "mentor" was most frequently used to describe the relationship between staff and interns, much of the initiation of new hires paralleled that strategy. Most employers reported assigning newly hired engineers to a lead staff or manager. "When a new engineer comes on, we assign (them to) a senior engineer or principal engineer and they work together so the senior can review their work and teach the process."

Some also provided in-house training, mostly focused on company procedures, but relied on mentors to explain and supervise technical expectations. "All technical products prepared by people who have not yet been licensed are reviewed, so that reviewer is responsible for mentoring. It's part of their duties to make that investment in our future. *And we evaluate senior employees on their mentoring*."

Diversify Assignments

New hires are often deliberately assigned to diverse projects to provide cross-training in multiple departments and functions. They commonly rotate among experienced engineers, directly paired or as a team, for active mentoring and guidance. This critical instruction is provided as on-the-job training. For many employers, the approach is not entirely new, but has become more deliberate and strategic than in the past. Many senior engineers reported doing assignment rotation early in their careers, but remembered their experiences as involving a smaller number of people. Now firms are trying to increase the exposure of new engineers to projects across the organization, both to expand the scope of the recruit's experience but also to utilize the best mentors and instructors, regardless of their specialty.

Only a few of these programs were formally defined. Most were impromptu assignments. In addition to providing a mentor and exposure to a variety of assignments, some move the new hire gradually from simple to more complex projects. "Typically, we try to put new engineers in new service area groups so they can start with residential and small commercial buildings. Then, once they have more experience, they move to where they can get more experience. (It's) not a program, though, just do it on an individual basis."

In some cases, this stair-stepping approach to increasing responsibility is done instead of mentoring, with the supervisor of each unit responsible for the new engineer's work. "We tend to give them smaller projects to start with. Don't directly assign them a mentor, although they are typically mentored with senior engineers since they usually don't have a license so the senior engineer has to sign off on their designs. Designs are reviewed as they go through a project with more senior engineers who are going to stamp the design and sign off, and then many of our sites are dual purpose so we generally train by doing small projects that develop into larger ones."

9.3.5 *Retention and Retraining Issues*

The retention of new hires through mid-career engineers has been a consistent issue for the industry, although it waned somewhat during the recession. Indications are that the industry may lose 10–15 % of its engineers per year for non-retirement reasons. Reasons for this include difficulty keeping salaries competitive and challenges due to work location.

While a few utilities have found themselves able to "sell" the advantages of small town and rural life as benefits, others have found that engineers recruited from urban areas often wish to return to them. Even utilities in cities such as Tacoma reported that younger engineers often left after only a few years, preferring urban Seattle or returning to the East Coast.

Interestingly, several employers noted that former interns were more likely to stay, reflecting the substantial benefit that resulted from candidates who knew more about the organizations—and the lifestyles—to which they were committing.

9.3.5.1 Supplementing the Skills of Experienced Workers

Finding the most effective way to nurture and sustain mid-career engineers will continue to be critical retention and development issues as senior engineers retire.

The challenge of replacing mid-career engineers was mentioned by many of the employers we interviewed. Cost was one issue, but so was simply the length of time it takes to gain relevant experience. Many of these employees have specialized experience in a critical aspect of the industry, but lack well-rounded, mid-level skills and experience. One employer noted: "(These positions) can be very hard to fill because people with those skills aren't readily available. They're highly-skilled specialists."

> "We can't hire all newbies just out of college. What we really need is more in the middle, experienced people, but we just can't find them. We're already stretched. We can't bring a bunch of new people on board and train them up. It's especially hard in key technical positions to hire for experience."
>
> Pacific Northwest employer

Upgrades and Mid-career Training

Mid-career training is important to retain experienced power engineers and also to compensate for the loss of institutional memory due to the retirement of senior engineers.

Employers acknowledged that mid-career engineers were often lacking computer skills—from programming to basic computer user skills—that were sometimes were more common among younger, less experienced engineers. One employer noted, "Advanced software skills, database and programming skills; pulling data into Excel, presenting the data to non-technical audiences, report writing and technical documentation—(this is often) harder for senior engineers than juniors."

This may seem contradictory at first, but while employers often found new engineers lacking in some technical skills, particularly those related to explaining and presenting technical materials to a general audience, they found mid-level staff had not kept up with computer-related developments. Their skills too often were underdeveloped or outdated.

Another critical aspect of upgrade training is compensating for the loss of institutional memory due to the departure of senior engineers. As one employer noted:

> Employees may have experience at another utility, but we are lacking someone with 20–30 years of experience with *us*. We had an early-out package in the 1990s and lost some of our most senior engineers. Institutional knowledge went with them. That was worsened by turnover in some groups, like protection and control, and low tenure resulted.

These losses have caused employers to redouble efforts to retain mid-level staff, even while "poaching" by other employers—and sometimes even among their own departments—increases.

In-House Training Programs

As a result of these conditions and because some larger firms have managed to retain formal training programs, a few employers reported providing structured in-house training to address common employee development needs. These internal trainings were frequently described in terms of generic skills, such as communication and planning skills. "(We) also have an (internal) offering 6–12 times per year, sponsored by the civil side of the house, to talk about project management areas (open for all)."

But even these companies noted the current budget constraints made training difficult to achieve. Some noted that training was hampered not by a lack of a training budget, but because they were too short-staffed to allow time for training. One manager noted, "What drives engineers to learn is having a problem to solve, so you do that from experience and with mentoring. If there are skills they are lacking, then we will look for that, for a specific technical skill (to provide)." However, those specific skills were generally gained on the job.

The lingering effects of budget constraints meant employers were looking to minimize training costs, even while many recognized a need for continuous growth by this valuable group of employees. A few tried to give employees time to learn, rather than providing training. "Essentially we provide on-the-job training... We don't invest a lot in long-term training, but sometimes in short-term, on a case-by-case basis. It's not so much skills development as exploring a new knowledge area. We'll often allocate 'thinking time,' a 3- to 6-months period to try out a new idea." As may be expected, these practices were most common among employers who conduct the most research and development.

All employers also reported relying on external training providers, such as vendors providing support for new equipment. A few reported supporting employees to attend external courses, including a few who sponsored employees to attend power system courses outside the region. Most often, this practice was mentioned as part of the pursuit of employee promotion and project requirements. "For example, we had a very knowledgeable engineer on the generation side, who needed better modeling skills, so that employee attended training on power software and built those skills. That was our need, driven by compliance."

Employers also mentioned the specific skills most needed by mid-career and senior staff, as compared to those needed by new hires.

> Change management, which seems hard on older engineers, learning to accept doing things differently.
>
> Technical writing seems to be a bigger problem for experienced engineers than new hires.

9.3.6 Attracting Youth to the Industry

All of the employers interviewed for this study had some form of internship and working relationships with the engineering departments at the state's four largest public and private engineering schools. Discussion of internships naturally led to discussion of the longer pipeline and the exposure of younger students, including high school students, to the industry. One employer noted:

> One of the reasons we sponsor internships is students often think the power industry is a dying industry. Juniors (college interns) often say 'a project I worked on was beneficial to changing my mind about the industry.' I know the line sections (line workers) demonstrate to high schools what the power industry is about, but I'm not sure how we do. Need to think about it more.

Several employers commented on the power of internships and student outreach to change the image of the industry. The former persona of the industry as being staid and uninteresting is being replaced by one that promises new learning experiences and the satisfaction of working in a challenging, high-tech environment in a specialty that also "makes a difference" to society. One employer added:

> From a generation side, if we could somehow convey the diversity of the work. They really do get a lot of job satisfaction because you may work on fish rearing 1 day, hydro units the next day, the sites are all beautiful to go to, you have the potential to do automation with computer programming and cyber security—there's always something to learn, a lot of job satisfaction, and there's just a lot of things you can do, you never get bored, and trying to convey that to students—I'm not sure how to do it other than bringing them here to see what they get to work with.

Student outreach to local secondary schools also provides opportunities for employers to target engineering as well as the full range of careers available in the electric power industry. Field trips and projects connect students with employees in the workplace, and also serve as a vehicle for emphasizing the importance of academic preparation in science, engineering, technology, and math (STEM) for energy careers. One employer noted:

> In (our area) there is a STEM high school that requires students, as part of their process, to find a role in an organization and do work assignments. We have had two for a full year. They worked with computer technicians because that was their interest, but we would do it again. We also do career fairs. Staff rotate responsibility with the local middle and high schools, usually twice a year.

Yet, for some organizations, budget constraints have limited how much outreach they are able to do. The impact of limited funding on student hiring for one large public employer was especially pronounced:

> We work to help high schools by talking about engineering as well as trade and craft occupations. We had also planned to hire local students for summer employment again this year, but the federal hiring freeze ended that. We have done it before. There used to just be three or four, but last summer I think we had 12. This year we were going to have 13.

9.4 Conclusions and Recommendations

The power industry has changed since the 2008 workforce study of electric power employers, and it continues to grow more complex. Some of the most critical workforce problems noted then, including looming retirements, stalled during the recession but did not go away. Other problems grew worse, as engineering departments faced the same economic cutbacks that undermined much of higher education, compounded by retirements among engineering faculty.

As a result, the industry still faces critical workforce challenges. Although electrical engineering enrollments appear to be up nationwide, new industries increasingly compete with traditional power generation for graduates. Employers report that only a small number of graduates have any experience with the power industry, perhaps least of all with traditional hydropower generation and distribution. And many engineering students are foreign nationals, who cannot meet security clearances required by some industry employers.

Yet the industry needs workers now and in the future. While retirements abated during the recession, the senior workers did not get younger while waiting for the economy to recover, and the potential for substantial loss of experience in the industry workforce remains. The data collected for this study shows that while employers project a modest number of new hires in the near future, at the time data was collected 85 power engineering vacancies were reported, most at the mid-levels of experience. Perhaps more important is that employers expect 210 retirements to occur among current power engineers by 2018, which is more than 20 % of the current power engineering workforce. It is worth noting that the study sample of 18 firms did not account for retirements at other organizations; thus, the results likely understate the actual number of retirements in Washington and across the region.

Further, jobs continue to grow more multifaceted, reflecting substantive change from smart grid installation and regulatory increases. These changes affect the sheer quantity of knowledge and skills required of new power engineers, and continue to increase skill demands among mid-career engineers. In particular, employers uniformly reported that new power engineers lack exposure to the industry and, thus, basic understanding of its requirements. They lack many basic workplace skills, including teamwork skills, basic workplace etiquette and business basics. Most are very familiar with computers and software, but not the elements of smart grid operation. Too few know how to generate and use data, and fewer can present data to a general audience. Yet these are considered core skills by every employer.

Mid-career power engineers, who are needed to step into senior positions as retirements occur, have too often developed specialized skills but not the more well-rounded proficiencies required for senior leadership. Frequently, they have underdeveloped computer and data skills, and not all welcome opportunities to broaden their range of skills.

Employers report utilizing a variety of methods to address these issues:

- Retaining senior staff through retire/rehire and emeritus options.
- Relying on contractors to fill temporary workforce needs.
- Utilizing college internships to attract engineering students and expose them to the industry.
- Using mentors and rotating assignments to initiate new engineers to the industry workplace and culture.
- Working with the K-12 system to attract younger students to the industry.

9.4.1 Recommendations

The employer data collected for this study shows that there is strong current demand for qualified power engineers, and that employers are also hiring at the entry-levels to fill current openings due to labor shortages. Although employers do not anticipate adding a large number of new power engineer positions over the next few years, a sizable number of current engineers are predicted to retire, and employers plan to replace all of those openings.

Because this study did not include all power industry employers, and because trained power engineers are also in demand by consulting organizations, technology companies, manufacturers, suppliers and other industry sectors, it is likely that the true future demand for power engineers in Washington and across the region will exceed the totals identified in this study. Therefore, one recommendation is that Washington State University, as well as other regional institutions, should continue to develop, enhance and expand power engineering program capacity so that the supply of trained engineers is adequate to meet current and future demand in the power industry and in affiliated industry sectors in Washington and across the region. Periodic surveys of regional employers should be conducted to help confirm hiring trends and match labor market supply and demand.

A second recommendation is that existing power engineering programs should be reviewed to ensure that students are engaged in learning experiences that impart the knowledge, skills and abilities that employers require, including the technical and non-technical workforce skills that students will need to succeed in the workplace. The review process should include discussion and content review with industry staff, many of whom expressed strong opinions about what current programs lack; many employers agreed to participate in these interviews because they wanted their views to reach educators, which signals a willingness to also participate more formally in work to update curricula.

A third recommendation is that students should be encouraged—or even required—to participate in internships and other industry-based experiences that allow them to apply their academic learning in a work setting, and which help to equip students with an understanding of the industry culture and employer expectations that is essential for career success. Many employers already offer work-based learning experiences to power engineering students, and more employers are likely willing to provide these opportunities, which offer myriad, mutual benefits to students, employers, and institutions alike.

Finally, regional universities should consider the implications of future technology enhancements and power engineering workforce retirements for the opportunities these transitions may present. Mid-career professionals will require skill upgrades as they design, implement, and adapt to new technologies, and as they prepare to replace senior engineers who retire. This could present new opportunities for regional universities to provide graduate-level training, continuing education and professional certification, leadership development, and research services to the power industry.

Acknowledgements The authors would like to thank the many industry participants and supporters of this project. Special thanks go to the employers and survey respondents who provided data and agreed to be interviewed for this study. Their many contributions to this research are greatly appreciated. The authors are grateful to Jody Opheim of the WSU-ESIC for their sponsorship, and to Sally Zeiger Hanson and Melinda Spencer (WSU Energy Program) for content contributions, draft reviews, and editing support.

References

1. See: Hardcastle A (2008) Workforce challenges of electric sector employers in Washington and Oregon. Prepared by the Washington State University Energy Program for the Center of Excellence in Energy Technology (now Pacific Northwest Center of Excellence for Clean Energy-Centralia College): http://www.energy.wsu.edu/Documents/WSU_Workforce_Challenges_Final_Report_090311.pdf
2. Hardcastle A, Jull P, Zeiger Hanson S (2013) Workforce challenges of electric power employers in the Pacific Northwest. Washington State University Energy Program, for the Pacific Northwest Center of Excellence in Clean Energy. See: http://cleanenergyexcellence.org/resources/
3. Hinkel JR (2008) The 2003 Northeast blackout – five years later, Scientific American. Accessed 13 Aug 2008
4. National Science Foundation (2008) Workshop on the future power and energy workforce. 29–30 Nov 2007, Arlington, VA. http://ecpe.ece.iastate.edu/nsfws/. Accessed 8 Sept 2008
5. U.S. Power and Energy Engineering Workforce Collaborative (2009) See: Preparing the U.S. Foundation for future electric energy systems: a strong power and energy engineering workforce. IEEE-Power Energ Soc. See: http://www.ieee-pes.org/images/pdf/US_Power_&_Energy_Collaborative_Action_Plan_April_2009_Adobe72.pdf
6. Center for Energy Workforce Development. Gaps in the energy workforce pipeline: 2008 CEWD survey results. http://www.cewd.org/documents/CEWD_08Results.pdf. The study investigated a wide range of power industry technical job categories, including lineworkers, pipefitters, pipelayers, engineers, plant operators, and technicians
7. Center for Energy Workforce Development. Gaps in the energy workforce pipeline, 2011 CEWD survey results. http://www.cewd.org/surveyreport/CEWD-2011surveyreport-021512.pdf
8. See: Hardcastle A, Jull P, Zeiger Hanson S (2013) Workforce challenges of electric power employers in the Pacific Northwest. Washington State University Energy Program, for the Pacific Northwest Center of Excellence in Clean Energy. See: http://cleanenergyexcellence.org/resources/
9. U.S. Power and Energy Engineering Workforce Collaborative (2009) Preparing the U.S. Foundation for future electric energy systems: a strong power and energy engineering workforce. IEEE-Power Energ Soc. See: http://www.ieee-pes.org/images/pdf/US_Power_&_Energy_Collaborative_Action_Plan_April_2009_Adobe72.pdf
10. See: Hardcastle A, Jull P, Zeiger Hanson S (2013) Workforce challenges of electric power employers in the Pacific Northwest. In 2011 the average annual wage for utilities (all employees) in Washington State was over $79,000. State labor market data shows that the average annual wage for electrical engineers (March 2012) was $93,967; however, these data do not report specifically for power engineers. See: https://fortress.wa.gov/esd/employmentdata/reports-publications/occupational-reports/occupations-in-demand

MIX
Papier aus verantwortungsvollen Quellen
Paper from responsible sources
FSC® C105338

If you have any concerns about our products,
you can contact us on
ProductSafety@springernature.com

In case Publisher is established outside the EU,
the EU authorized representative is:
Springer Nature Customer Service Center GmbH
Europaplatz 3, 69115 Heidelberg, Germany

Printed by Libri Plureos GmbH
in Hamburg, Germany